PENGUIN CLASSICS

THE LOG FROM THE *SEA OF CORTEZ*

Born in Salinas, California, in 1902, John Steinbeck grew up in a fertile agricultural valley about twenty-five miles from the Pacific Coast—and both valley and coast would serve as settings for some of his best fiction. In 1919 he went to Stanford University, where he intermittently enrolled in literature and writing courses until he left in 1925 without taking a degree. During the next five years he supported himself as a laborer and journalist in New York City, all the time working on his first novel, *Cup of Gold* (1929). After marriage and a move to Pacific Grove, he published two California books, *The Pastures of Heaven* (1932) and *To a God Unknown* (1933), and worked on short stories later collected in *The Long Valley* (1938). Popular success and financial security came only with *Tortilla Flat* (1935), stories about Monterey's paisanos. A ceaseless experimenter throughout his career, Steinbeck changed courses regularly. Three powerful novels of the late 1930s focused on the California laboring class: *In Dubious Battle* (1936), *Of Mice and Men* (1937), and the book considered by many his finest, *The Grapes of Wrath* (1939). Early in the 1940s, Steinbeck became a filmmaker with *The Forgotten Village* (1941) and a serious student of marine biology with *Sea of Cortez* (1941). He devoted his services to the war, writing *Bombs Away* (1942) and the controversial play-novelette *The Moon Is Down* (1942). *Cannery Row* (1945), *The Wayward Bus* (1948), another experimental drama, *Burning Bright* (1950), and *The Log from the* Sea of Cortez (1951) preceded publication of the monumental *East of Eden* (1952), an ambitious saga of the Salinas Valley and his own family's history. The last decades of his life were spent in New York City and Sag Harbor with his third wife, with whom he traveled widely. Later books include *Sweet Thursday* (1954), *The Short Reign of Pippin IV: A Fabrication* (1957), *Once There Was a War* (1958), *The Winter of Our Discontent* (1961), *Travels with Charley in Search of America* (1962), *America and Americans* (1966), and the posthumously published *Journal of a Novel: The* East of Eden *Letters* (1969), *Viva Zapata!* (1975), *The Acts of King Arthur and His Noble Knights* (1976), and *Working Days: The Journals of* The Grapes of Wrath (1989). He died in 1968, having won a Nobel Prize in 1962.

Richard Astro is professor of English at the University of Central Florida, where he is also director of the Eastern Europe Linkage Institute. He is the author of *John Steinbeck and Edward F. Ricketts: The Shaping of a Novelist,* as well as studies on Hemingway, Fitzgerald, and western American literature.

THE LOG FROM THE *SEA OF CORTEZ*

JOHN STEINBECK

The narrative portion of the book, *Sea of Cortez* (1941),
by John Steinbeck and E. F. Ricketts

INTRODUCTION BY
RICHARD ASTRO

PENGUIN BOOKS

PENGUIN BOOKS
Published by the Penguin Group
Penguin Group (USA) Inc., 375 Hudson Street, New York, New York 10014, U.S.A.
Penguin Group (Canada), 90 Eglinton Avenue East, Suite 700, Toronto,
Ontario, Canada M4P 2Y3 (a division of Pearson Penguin Canada Inc.)
Penguin Books Ltd, 80 Strand, London WC2R 0RL, England
Penguin Ireland, 25 St Stephen's Green, Dublin 2, Ireland (a division of Penguin Books Ltd)
Penguin Group (Australia), 250 Camberwell Road, Camberwell,
Victoria 3124, Australia (a division of Pearson Australia Group Pty Ltd)
Penguin Books India Pvt Ltd, 11 Community Centre, Panchsheel Park, New Delhi – 110 017, India
Penguin Group (NZ), cnr Airborne and Rosedale Roads,
Albany, Auckland 1310, New Zealand (a division of Pearson New Zealand Ltd)
Penguin Books (South Africa) (Pty) Ltd, 24 Sturdee Avenue,
Rosebank, Johannesburg 2196, South Africa

Penguin Books Ltd, Registered Offices: 80 Strand, London WC2R 0RL, Englan

Sea of Cortez first published in the United States of America
by The Viking Press 1941
The Log from the Sea of Cortez first published by The Viking Press 1951
Published in Penguin Books 1977
This edition with an introduction by Richard Astro
published in Penguin Books 1995

30 29 28 27 26 25 24 23 22 21

LIBRARY OF CONGRESS CATALOGING-IN-PUBLICATION DATA
Steinbeck, John, 1902–1968.
[Sea of Cortez]
The log from the Sea of Cortez/John Steinbeck; introduction by
Richard Astro.
p. cm.
"The narrative portion of the book, Sea of Cortez (1941), by John
Steinbeck and E. F. Ricketts."
Originally published: New York: Viking, 1951.
"Appendix: About Ed Ricketts": p.
Includes bibliographical references and index.
ISBN 0 14 01.8744 8 (pbk.)
1. Marine invertebrates—Mexico—California, Gulf of.
2. California, Gulf of (Mexico)—Description and travel.
3. Steinbeck, John, 1902–1968—Journeys—Mexico—California, Gulf
of. 4. Ricketts, Edward Flanders, 1896–1948—Journeys—Mexico—
California, Gulf of. I. Ricketts, Edward Flanders, 1896–1948.
II. Steinbeck, John, 1902–1968. About Ed Ricketts. III. Title.
QL225.S74 1995
508.3164'1—dc20 95-14802

Printed in the United States of America
Set in Stempel Garamond

CONTENTS

INTRODUCTION

In February 1995, a large and diverse group of Californians, most of them at least in their mid-seventies, gathered on Cannery Row to celebrate the fiftieth anniversary of the publication of Steinbeck's novel of the same name, and otherwise to reminisce about the two men who made the Row famous: the novelist himself and his closest personal and intellectual companion, marine biologist Edward F. Ricketts. The event was billed as "a symposium," and was co-sponsored by the Cannery Row Foundation and Steinbeck Research Center at San Jose State University. But given the list of participants—including two of Ricketts's children; Joel Hedgpeth, senior curmudgeon of the California intertidal; Virginia Scardigli, former teacher and friend of both Steinbeck and Ricketts; Alan Baldrige, for many years the librarian at Stanford University's Hopkins Marine Station on Ocean Avenue near the Row; and Robert Enea, a nephew of two of the crew members from the Sea of Cortez expedition—the event was less a symposium than a giant party. And this seemed an appropriate way to commemorate the publication of the book in which Steinbeck wrote that every party has its own pathology, and that "a party hardly ever goes the way it is planned or intended." Of course, that book's leading character is a fictionalized version of Steinbeck's closest friend and his collaborator on *Sea of Cortez*—his most important work of nonfiction, a volume which contains the core of Steinbeck's worldview, his philosophy of life, and the essence of a relationship between a novelist and a scientist that ranks among the most famous friendships in American letters. If many tall tales were told at the symposium, embellished by years of telling, it made no difference, except to enhance the festivities. For whatever the excesses, the surviving few from the Steinbeck-Ricketts years knew and talked about the breadth and depth of a friendship that was deep and permanent, and that, because of the impact of Ricketts's thinking on Steinbeck's most important fiction, accounts in large measure for the novelist's success as a writer.

Cannery Row was published five years after the Steinbeck-Ricketts expedition to the Gulf of California, and while Ricketts's life in Monterey remained largely unchanged afterward (he was drafted into the army during World War II, but never left the Monterey presidio), Steinbeck departed California altogether. His marriage to his first wife, Carol, ended. He romanced Hollywood singer Gwen Conger, married her in New Orleans, joined the war effort as a correspondent for the New York *Herald Tribune*, wrote a novelette about the war entitled *The Moon Is Down* (1942) and some propaganda pieces for the Army Air Corps that were later published as *Bombs Away* (1943), bought a brownstone on Manhattan's East Side, and gradually became a New Yorker. He and Ricketts communicated by mail, but they hardly ever saw each other again.

Cannery Row, which Steinbeck claims he wrote for a group of soldiers who told him to write something funny, something that wasn't about the war, is more nostalgia than anything else, and the leading character, Doc, is a Ricketts who sometimes resembles the original and is at other times purely a creation of Steinbeck's imagination. He is not the Ricketts who co-authored *Sea of Cortez*, which was published days before Pearl Harbor was bombed and America entered the war that separated two men whose ideas were so closely interrelated that it is sometimes difficult to know who learned what from whom. That relationship and the thinking of the two men who wrote it are what *Sea of Cortez* is really all about. It is a useful work of travel literature, and it is a pioneering work of intertidal ecology, though it was written a full three decades before Earth Day turned environmental thinking into one of our national pastimes.

When Steinbeck died in December 1968, his critical reputation as a writer was severely tarnished. He had written little of significance in nearly two decades, and his support of the American war effort in Vietnam had put him in critical disrepute among even those critics who earlier had commended him as the champion of the victims of the Oklahoma dustbowl and the avarice of California agribusiness in *The Grapes of Wrath*, and for his compelling portraits of the simple but decent denizens of the Central California valleys in *Of Mice and Men*, *The Red Pony*, and *The Pastures of Heaven*. When he died, there were few serious scholars who did not share Harry T. Moore's feeling that his ultimate status as a writer would be that of a Louis Bromfield or a Bess Streeter Aldrich, and that even his best books were watered down by what Arthur Mizener called his "tenth-rate philosophizing."

History has proved otherwise. During the past quarter century, a veritable Steinbeck industry has emerged. All of his books have been reprinted. Important full-length critical studies have been published by major academic presses, and articles on virtually every aspect of his work have appeared in the best scholarly journals. The publication of his letters by his widow, Elaine, in collaboration with Robert Walsten, and a comprehensive and carefully researched biography by Jackson J. Benson, have shed new light on the man and his creative process. Steinbeck research centers now exist at several universities, most notably in the unlikely location of Muncie, Indiana, where, at Ball State University, Tetsumaro Hayashi began in 1969 publishing the *Steinbeck Quarterly*, which helped young Steinbeck scholars to share their views long before the more prestigious journals were prepared to question the judgments of Harry Moore and Arthur Mizener.

Today, Steinbeck's reputation seems secure. While few would disagree that his canon as a whole reflects an uneven talent, it is clear that his best books champion ordinary men and women, simple souls who do battle against the forces that dehumanize the species, and who struggle, sometimes successfully, sometimes not, to forge lives of genuine meaning and worth. At the center of Steinbeck's thematic vision is a continuing dialectic between contrasting ways of life: between innocence and experience, between primitivism and progress, between narrow self-interest and an enduring commitment to the human community. His most interesting characters—George Milton and Lennie Small in *Of Mice and Men*, Doc Burton of *In Dubious Battle*, Tom Joad and Jim Casy in *The Grapes of Wrath*, and Mack and the boys in *Cannery Row*—search for meaning in a world of human error and imperfection.

At the heart of this dialectic are the contrasting views of human society held by the novelist and Ed Ricketts. This contrast in views can be seen in *Sea of Cortez*, and in large measure accounts for the book's importance. For while in much of his work, and most notably in *The Grapes of Wrath*, Steinbeck celebrates what he calls "man's proven capacity for greatness of heart and spirit," the fact that man "grows beyond his work, walks up the stairs of his concepts, emerges ahead of his accomplishments," he also concedes (in the narrative portion of *Sea of Cortez*) that man "might be described fairly adequately, if simply, as a two-legged paradox. He has never become accustomed to the tragic miracle of consciousness. Perhaps, as has been suggested, his species is not set, has not jelled, but is still in a state of becoming, bound by his physical memories to a past of struggle and survival, limited in his futures by the uneasiness of thought and consciousness."

I have long believed and I have written elsewhere that "the tragic miracle of consciousness" is, for Steinbeck, man's greatest burden and his greatest glory. And it is the manner in which Steinbeck portrays this burden and this glory in his novels and his short stories that accounts for his success as a writer. This is the basis of the feeling in his fiction, the compassion, and at its extreme, his sentimentality. It was his central concern as a writer, from Henry Morgan's drive for power in *Cup of Gold* and Joseph Wayne's search for meaning in *To a God Unknown*, to the last sentence of his Nobel Prize acceptance speech, in which he paraphrased John the Apostle, stating, "In the end is the Word, and the Word is Man, and the Word is with Man." It is to *Sea of Cortez* that we must look if we are fully to understand all this, if we are to grasp the thematic vision of this writer whose books continue to be read and reread by millions of all ages, in his native California, across the United States, and throughout the world—where, in such diverse countries as Portugal and Poland, Mexico and Moldova, Steinbeck remains among the most loved and appreciated of all American novelists.

Though Steinbeck was born and grew up in the city of Salinas, a major processing center for the foodstuffs raised in one the most fertile agricultural lands in America, he spent much of his childhood and adolescence in the towns along nearby Monterey Bay. In 1930, he settled in the bayside community of Pacific Grove with his bride, Carol Henning, whom he met and married in nearby San Jose. The center of California's sardine fishing industry, Pacific Grove and its neighboring communities of Monterey and Carmel were for many years California's "seacoast of bohemia." Robinson Jeffers built Tor House along Big Sur. Robert Louis Stevenson, Jack London, and Ambrose Bierce were frequent short-term visitors, and Charles Warren Stoddard, George Sterling, and Mary Austin were permanent residents. Monterey Bay itself, as Robert Louis Stevenson wrote in "The Old Pacific Capital," resembles a giant fishhook—with Monterey cozily ensconced beside the barb. Just outside the barb, in a cove embraced by rugged Point Lobos, lies Carmel. And just short of Point Lobos, the Carmel River reaches the sea, flowing down from what Stevenson called "a true California valley, bare, dotted with chaparral, overlooked by quaint, unfinished hills."

The Steinbecks were in poor financial shape as the decade began. His first novel, *Cup of Gold*, failed to sell, and Carol had given up a teaching job in San Jose to move with him to the Steinbeck cottage in Pacific Grove. When Steinbeck and Ricketts met in 1930 (not at a dentist's office as Steinbeck states in his retrospective "About Ed Ricketts," but rather at the home of Ricketts's friend and other collaborator, Jack Calvin), the most immediate result of their budding friendship was that Ricketts hired Carol as his secretary at his Pacific Biological Laboratory, where Ricketts made ends meet during the Great Depression by selling prepared slides to local high schools. At the same time, Steinbeck and Ricketts gradually developed a deep and lasting friendship, based largely on the novelist's interest in Ricketts's work in the intertidal.

It is generally assumed that Steinbeck's interest in marine science began when he met Ricketts. But Steinbeck had been interested in the subject for several years, at least since 1923, when he took a summer course in general zoology at the Hopkins Marine Station taught by C. V. Taylor. Taylor was a student of Charles Kofoid at Berkeley, and both were devotees of William Emerson Ritter, whose doctrine of the organismal conception of life formed the zeitgeist of the Berkeley biological sciences faculty at the time. In fact, Ritter's ideas were transmitted via Kofoid and Taylor to the young and impressionable Steinbeck, who years later told Hopkins professor Rolf Bolin that what he remembered most about his summer at Hopkins was Ritter's concept of the "superorganism."

Ritter believed that "in all parts of nature and in nature itself as one gigantic whole, wholes are so related to their parts that not only does the existence of the whole depend upon the orderly cooperation and interdependence of the parts, but the whole exercises a measure of determinative control over its parts." This notion of "wholeness" is inherent in every unit of existence, claimed Ritter, since each living unit is a unique whole, the parts of which "contribute their proper share to the structure and the functioning of the whole." Ritter believed that since "one's ability to construct his own nature from portions of nature in general is a basic fact of his reality," man is capable of understanding the organismal unity of life and, as a result, can know himself more fully. This, says Ritter, is "man's supreme glory"—not only "that he can know the world, but he can know himself as a knower of the world."

Ed Ricketts was not familiar with Ritter's work when he came to California in 1923, after an uneven career as a biology undergraduate at the University of Chicago (he grew up on the northwest side of the city). But Ritter's ideas had much in common with those of Ricketts's favorite teacher at the university, animal ecologist W. C. Allee, whose ideas about the universality of social behavior among animals, and whose theory that animals behave differently in groups than as individuals (described in detail in his classic 1931 treatise on the subject, *Animal Aggregations*), profoundly affected Ricketts's way of viewing life. Years later, Jack Calvin told this writer that "we knew W. C. Allee from Ed's conversations, discovering that all of his former students got a holy look in their eyes at the mention of his name, as Ed always did." Allee did much of his work at Woods Hole, Massachusetts, where he eventually concluded that "the social medium is the condition necessary to the conservation and renewal of life," but that this is an automatic and not a conscious process. And when Allee turned his attention from the lower animals to man, he concluded that so-called altruistic drives in man "apparently are the development of these innate tendencies toward cooperation, which find their early physiological expression in many simpler animals."

Ritter's organismal conception, his idea that the whole is more than the sum of its parts and that these parts arise from a differentiation of the whole, is different from but complementary to Allee's thesis that organisms cooperate with one another to ensure their own survival. The ideas of these two pioneering ecologists provided an expansive intellectual ground upon which Steinbeck and Ricketts could develop their friendship. From almost the first day of their meeting, they became members of a larger group of latter-day Cannery Row bohemians, bound together by their poverty, which they combatted, as Jack Calvin noted, "by raiding local gardens and stealing vegetables for communal stews." Over time, Steinbeck drew very close to Ricketts. They spent endless hours in Ed's lab discussing the work of Allee and Ritter as Steinbeck worked on his novels and short stories and Ricketts studied what he called "the good, kind, sane little animals," the marine invertebrates of the Central California coast.

In time, they both succeeded. Steinbeck achieved modest successes with his early short stories, greater glory with *Tortilla Flat*, which won him critical recognition, and then—when he sold the movie rights to the novel for the then-magnificent sum of four thousand dollars—financial independence. In the late 1930s, his popularity skyrocketed as *Of Mice and Men* succeeded both as fiction and as theater, and as *In Dubious Battle* and *The Grapes of Wrath* established him as a champion of the proletariat. *Grapes* was and remains Steinbeck's masterpiece. This epic account of the plight of a family of disinherited Oklahoma tenant farmers made Steinbeck a novelist of international stature. It is the book upon which his enduring reputation as a major American writer continues to rest.

Ricketts, on the other hand, worked away on his studies of life in the tidepools, taking the necessary time to maintain his prepared-slide business, which was his only source of income until 1939. That year, Stanford University Press published the results of his work in *Between Pacific Tides*, which Ricketts co-authored with Jack Calvin. Calvin did little more than polish Ricketts's stilted prose into a thoroughly readable and very professional account of the habits and habitats of the animals living on the rocky shores and in the tide pools of the Pacific Coast. Some years later, Steinbeck wrote a foreword to the third edition of *Tides*, noting that the book "is designed more to stir curiosity than to answer questions. . . . There are good things to see in the tidepools and interesting thoughts to be generated from the seeing. Every new eye applied to the peephole which looks out at the world may fish in some new beauty and some new pattern, and the world of the human mind must be enriched by such fishing." Ricketts's years of hard work paid off. *Between Pacific Tides* became the definitive sourcebook for studying marine life along the Pacific Coast, and even today it is read by students at every major oceanographic station from Southern California to British Columbia.

The Grapes of Wrath and *Between Pacific Tides* were both published in 1939. Both authors were left fatigued. Steinbeck had moved to the Los Gatos hills some two or three years earlier, but the two remained close friends and saw one another often. For some time they had planned to write a book together—originally, a modest handbook for general readers about the marine life of San Francisco Bay. Ricketts drafted an outline for the book, and Steinbeck (whose participation in the project has been largely unnoticed) suggested "shopping" the book to his publisher (Viking) and to Ricketts's (Stanford University), and giving it to the highest bidder. The book was to be written chiefly by Steinbeck, said Ricketts, and would be designed "so that it can be used by the sea coast wanderer who finds interest in the little bugs and would like to know what they are and how they live. Its treatment will revolt against the theory that only the dull is accurate and only the tiresome, valuable."

Even though Steinbeck wrote a three-thousand-word preface, and Ricketts over five thousand words of text, the Bay area handbook was never completed. It did, however, provide impetus for a larger, more expansive project, the 1940 collecting expedition to the Gulf of California which resulted in the subsequent collaboration on *Sea of Cortez: A Leisurely Journal of Travel and Research*. In addition to those members of the crew who are mentioned in the volume, Steinbeck's wife Carol made the trip, which the couple hoped would serve to help salvage a failing marriage. It didn't. The *Western Flyer* left Monterey Bay on March 11, and returned six weeks later on April 20. The four-thousand-mile trip covered some twenty-five to thirty collecting stations where Ricketts, Steinbeck, and the crew collected what Ricketts guessed was "the greatest lot of specimens ever to have been collected in the Gulf by any single expedition."

After the trip, Steinbeck and Carol returned to their home in Los Gatos, where their marriage promptly collapsed, and where Steinbeck was dragged into controversy over *The Grapes of Wrath*, which, during his absence, had been brutally attacked for its alleged communist sympathies. Typical was the charge by Phillip Bancroft of the Associated Farmers of California (and a former candidate for the United States Senate) that the novel "is straight revolutionary propaganda. . . . In page after page it tries to build class hatred, contempt for officers of the law, and contempt for religion." Steinbeck felt some vindication, however, when he learned in early May that *Grapes* had been awarded the Pulitzer Prize for fiction, though he was typically reticent about receiving the award, and turned over his one thousand dollars in prize money to a struggling Monterey writer named Richie Lovejoy, whose father had loaned Steinbeck money to begin his career a decade earlier.

Ricketts spent the better part of a year identifying and cataloging specimens, and many more months passed as the Viking Press assembled the volume, reproduced photographs of the most important animals collected, and dealt with the many criticisms and revisions of the authors as the book went to press. When Steinbeck returned to Cannery Row in January 1941, his marriage to Carol was over, and he was in the midst of a flourishing affair with singer Gwen Conger. He worked on the book's narrative, and with Ricketts on matters relating to its publication, throughout the spring and summer of 1941. Pascal Covici, Steinbeck's editor at Viking, probably spent more time on the publication of *Sea of Cortez* than on any three of Steinbeck's other books combined. It was finally published during the first week of December 1941. But the reviews in the papers of Sunday, December 7, were hardly noticed as readers were distracted by events of much more immediate importance.

Those reviews that did appear were mixed, but largely favorable. The venerable Clifton Fadiman was miffed. He was at a loss to understand how the author of *The Grapes of Wrath* got mixed up with such a project in the first place, and he and others pointed to parts of the narrative that seemed obscure, almost unreadable. Joseph Henry Jackson, then the arbiter of literary taste in San Francisco, thought it "suspicious mysticism." In terms of its scientific value, the critical response was more favorable. Among the more disparaging was that of John Lyman, who noted that the authors said a great deal about the "Panamic" character of the Gulf's fauna, but gave "only the bare lists of forms taken at each collecting station." More approvingly, Rolf Bolin, the Hopkins ichthyologist and longtime friend of Steinbeck and Ricketts, wrote that it was a good book and would be a great aid to people going to the area to collect. But whatever its scientific merits, the fact is that the book is recognized by nearly all of Steinbeck's critics as a statement of his beliefs about man and the world; that, as Peter Lisca noted as early as 1958, it "stands to his work very much as *Death in the Afternoon* and *Green Hills of Africa* stand to that of Hemingway." Accordingly, it is essential to dispel myths about the book's authorship and to understand just how it was written.

Sea of Cortez is a big book, nearly six hundred pages long. For many years, it was assumed that Steinbeck wrote the first part, the narrative of the trip—published separately by Viking in 1951 as *The Log from the* Sea of Cortez—and that Ricketts authored the second part, a phyletic catalog describing the animals collected, prefaced by a series of notes on preparing specimens. At the same time, it was believed that the material for the narrative came from two journals, one kept by Steinbeck, the other by Ricketts. Both assumptions are inaccurate. There were two journals, but neither was kept by Steinbeck. Rather, they were kept by Ricketts and by Tony Berry, the owner and captain of the purse seiner which Steinbeck and Ricketts chartered for the trip. And while Steinbeck referred to Berry's log for matters of fact (chiefly dates and times), he composed the narrative chiefly from Ricketts's journal. Indeed, in a joint memorandum which the authors wrote to Covici in August 1941, they set the record straight:

> Originally a journal of the trip was to have been kept by both of us, but the record was found to be a natural expression of only one of us. This journal was subsequently used by the other chiefly as a reminder of what had actually taken place, but in several cases parts of the original field notes were incorporated into the final narrative, and in one case a large section was lifted verbatim from other unpublished work. This was then passed back to the other for comment, completion of certain chiefly technical details, and corrections. And then the correction was passed back again.

In this memorandum to Covici, the authors dismiss the notion that *Sea of Cortez* is two books. Instead, they insist, "the structure is a collaboration, but mostly shaped by John. The book is the result."

The phyletic catalog is a comprehensive and remarkably readable account of marine life in the gulf, though it is not as complete as *Between Pacific Tides*, because it is based on a single collecting trip rather than on a decade of study and research. What is unusual about it as a work of science, however, is that it focuses on common rather than on rare forms of marine life—since, note Ricketts and Steinbeck, they, "more than the total of all rare forms, [are] important in the biological economy." The *Log* portion of the book is a fascinating series of accounts of the lifestyle of the Indians of the gulf, and discussions of birth and death, navigation and history, and even the scientific method itself. Among the best sections are those in which the writers ridicule science that is cut off from the real concerns of human life. They label such scientists as "dryballs" who create out of their own crusted minds "a world wrinkled with formaldehyde." Above all, though, the *Log* is a celebration of the holistic vision the authors shared, and in accordance with their "reverence" for the ideas of Allee and Ritter, this is depicted in terms more mystical and intuitive than scientific. "It is a strange thing that most of the feeling we call religious," they note in one of the most compelling passages in the book, "most of the mystical outcrying, which is one of the prized and used and desired reactions of our species, is really the understanding and the attempt to say that man is related to the whole thing, related inextricably to all reality, known and unknowable." The narrative as a whole is the record of scientific discovery intermingled with explorations into philosophy, "bright with sun and wet with sea water," and "the whole crusted over with exploring thought."

In "About Ed Ricketts," Steinbeck recalls that "very many conclusions Ed and I worked out together through endless discussion and reading and observation and experiment." They had a game, he notes, "which we playfully called speculative metaphysics. It was a sport of lopping off a piece of observed reality and letting it move up through the speculative process like a tree growing tall and bushy. We observed with pleasure how the branches of thought grew away from the trunk of external reality." Indeed, notes Steinbeck, "we worked together, and so closely that I do not now know in some cases who started which line of speculation since the end thought was the product of both minds. I do not know whose thought it was."

The Log from the Sea of Cortez is an exercise in speculative metaphysics, grounded in the factual record of the trip itself, though even here simple facts like dates get mixed up. Consider, for example, that chapter 24 records events that occurred on April 3. Chapter 25 continues the narrative but is dated April 22, and chapter 26 is dated April 5. And remember that the *Western Flyer* returned to port on April 20.

There are entire sections where the thinking of both men coincide, and it is difficult if not impossible to distinguish the authorship of ideas. Typical of these sections are those about the scientific method, about seeing life whole, and about how the mind of the observer inevitably colors what is observed. Both Ricketts and Steinbeck were avid enthusiasts of the work of John Elof Boodin, who wrote in *Cosmic Evolution* (1925) that "the laws of thought are the laws of things" (the phrase is used verbatim in the *Log*), and that this law underpins the very notion of human creativity, since man and man alone can be a knower and can use his knowledge to understand the universe.

There are other sections of the *Log*, however, where research into the composition of the narrative reveals single authorship. The complex and controversial chapter on what the authors call "non-teleological" thinking was written almost entirely by Ricketts a decade before *Sea of Cortez* was published. Steinbeck enlisted Paul de Kruif to help market it and two of Ricketts's other essays ("The Philosophy of Breaking Through" and "A Spiritual Morphology of Poetry") to the editors of *Harpers*, but Ricketts's convoluted prose and his complicated thinking made this an exercise in futility. So, to provide a forum for Ricketts's ideas, and because he thought he could find a way to incorporate them into the *Log* that would be unobtrusive and consistent with the tone of the manuscript as a whole, Steinbeck included the twenty-page essay as "an Easter Sunday sermon." And there are other sections of the narrative, specifically those dealing with the patterns of tides and with something the authors call "sea-memory," that date back to a collecting trip Ricketts made with Jack Calvin and with the now-legendary comparative mythologist Joseph Campbell in the early 1930s.

Most important, however, are those passages of the *Log* in which Steinbeck and Ricketts work out their differences in their views of the world and man's role in it, for it is in these sections that we find clues to what is really going on in such important novels as *In Dubious Battle* and *The Grapes of Wrath*. There are those who believe that Steinbeck drew most if not all of his ideas from Ricketts. Indeed, Jack Calvin speaks for more than a few of Ricketts's friends when he suggests that "Ed was a reservoir for John to draw on . . . in Ed he found an endless source of material—or call it inspiration if you like—and used it hungrily." The fact is, however, that the intellectual relationship between Steinbeck and Ricketts was a very complicated affair. They disagreed on matters of intellectual substance almost as often as they agreed. Those agreements and disagreements can be found in the *Log*, and are worked out in fictional form in Steinbeck's most important novels.

Though Ricketts read widely and was extraordinarily knowledgeable, his worldview was narrow in that it was essentially Eastern and mystical. Indeed, what he called nonteleological or "is" thinking is essentially noncausal thinking. His major thirst in life was to see and to understand, which he defined as "breaking through" (a phrase he found in Robinson Jeffers's "Roan Stallion" and quoted in his "Spiritual Morphology of Poetry") to an understanding of what he called "the deep thing," where we can see and know, quoting from William Blake's "Visions of the Daughters of Albion," that "all that lives is holy." For Ricketts, the objective was what he called "a creative synthesis," an "emergent viewpoint," where by living into the whole one can know "it's right, it's alright, the good, the bad, whatever is."

Ricketts's doctrine of "breaking through" is the cornerstone of his worldview. And certainly Steinbeck shared his friend's passion for living deeply, seeing clearly, and viewing life whole. Steinbeck's work at Hopkins predisposed him to holistic thinking, which he embraced fully, and Blake's statement that "all that lives is holy" is quoted verbatim by Jim Casy in *The Grapes of Wrath* and is the basis for collective action by the Joad family as they move from the "I" to the "we" and become leaders of a movement to empower the lonely and displaced tenant farmers. But for Steinbeck, simply understanding the deep thing, the fundamental unity of life, is essentially a monistic approach that ignores common human needs and so is socially flawed. From Ricketts, Steinbeck learned to see life in scientific terms. His own reading of Ritter, and years of conversations with Ricketts, helped him see life in largely biological terms. Perhaps that is why so many of his most memorable characters are animal-like in thought and action. Tularecito in *The Pastures of Heaven*, Noah Joad in *The Grapes of Wrath*, assorted denizens of Cannery Row and Tortilla Flat, and, most significantly, Lennie Small in *Of Mice and Men*, have more in common with what Ricketts called "the good, kind sane little animals" of the intertidal than with physicians or philosophers. But while Steinbeck understood and was sensitive to human weakness, and while he sometimes envied the simple Indians of the Gulf of California—who, as he notes in the *Log*, may one day have a legend about their northern neighbors, that "great and godlike race that flew away in four-motored bombers to the accompaniment of exploding bombs, the voice of God calling them home"—he was not content to view the world with what he identified as simple "understanding-acceptance." Rather, for Steinbeck, man is a creature of earth, not a heaven-bound pilgrim, and the writer's most memorable characters are those who see life whole, and then act on the basis of that understanding, to "break through" to useful and purposeful social action.

The clearest picture of the differences between Steinbeck and Ricketts regarding the proper course of human action for those who can "break through" can be drawn from a short film script Steinbeck wrote during the composition of *Sea of Cortez*, and an essay Ricketts wrote in response. Steinbeck returned to Mexico for a short time during the summer of 1940 with filmmaker Herb Klein to make a study of disease in an isolated village; this study was made into a well-received documentary entitled *The Forgotten Village*. The script focuses on the initiative of a young boy, Juan Diego, who is outraged because a deadly microbial virus, which has polluted the village's water supply and has killed his brother and made his sister seriously ill, is being treated by witch doctors when real medical help is nearby. Juan Diego leaves the village to find the doctors of the Rural Health Service, who return with him to cure the problem. Noting that "changes in people are never quick," Steinbeck prophesies that, because of the Juan Diegos of Mexico, "the change will come, is coming; the long climb out of darkness. Already the people are learning, changing their lives, working, living in new ways."

After reading Steinbeck's text, Ricketts wrote an essay he called his "Thesis and Materials for a Script on Mexico"—actually an antiscript to Steinbeck's. In it, Ricketts noted that "the chief character in John's script is the Indian boy who becomes so imbued with the spirit of modern medical progress that he leaves the traditional way of his people to associate himself with the new thing."

> The working out of a script for the "other side" might correspondingly be achieved through the figure of some wise and mellow old man, who has long ago developed beyond the expediencies of economic drives and power drives, and to whom for guidance in adolescent troubles some grandchild comes. . . . A wise old man, present during the time of building a high speed road through a primitive community, appropriately might point out the evils of the encroaching mechanistic civilization to a young person.

In his best fiction, Steinbeck worked out the conflict between primitivism and progress, between his own view of the world and that of Ricketts—both of which were based, of course, on a scientific view of life organized around the concept of wholeness which is as spiritual as it is biological. And the Ed Ricketts characters in Steinbeck's fiction (they are several and are usually named "Doc") are those who are somehow cut off. They see and understand, but they cannot act on the basis of that understanding for the betterment of the species. Doc Burton in *In Dubious Battle* sees and understands the plight of the striking apple pickers in the Torgas Valley, but he wanders off into the night, frustrated by his inability to act on their behalf. He is "reincarnated" as Jim Casy in *The Grapes of Wrath*, who returns as Christ from the wilderness, and, seeing life whole, realizing that "all that lives is holy," gives his life to aid the dispossessed and disinherited. And there is Doc in *Cannery Row*, who wants only to "savor the hot taste of life," even as the Row itself (which for Doc and his friends is "a poem, a stink, a grating noise, a quality of light, a tone, a habit, a nostalgia, a dream") is really an island surrounded by an encroaching society which will ultimately destroy it. Little wonder the book is dedicated "to Ed Ricketts, who knows why or should." And there is its sequel, *Sweet Thursday*, where the Ricketts character seems even more isolated in a book which is less sweet than bittersweet. And finally there is that strange play-novelette, *Burning Bright*, in which the Ricketts character (named Friend Ed) teaches the Steinbeck character (Joe Saul) how to see and understand things whole and then how to receive (a trait which, in "About Ed Ricketts," Steinbeck identified as among Ricketts's greatest talents).

In the *Log*, Steinbeck writes a passage which could easily have been taken from the work of William Emerson Ritter (it appears nowhere in Ricketts's notes on the trip), in which he reflects that "there are colonies of pelagic tunicates which have a shape like the finger of a glove." Steinbeck remarks that "each member of the colony is an individual, but the colony is another individual animal, not at all like the sum of its individuals." And, says Steinbeck, "I am much more than the sum of my cells and, for all I know, they are much more than the division of me." There is "no quietism in such acceptance," notes the novelist, "but rather the basis for a far deeper understanding of us and our world." This is Ritter's organismal conception, which Steinbeck learned at Hopkins and discussed for so many years with Ricketts. At the core of the argument is the premise that, since given properties of parts are determined by or explained in terms of the whole, the whole is directive, is capable of directing the parts. In other words, the whole acts as a causal unit—on its own parts. As stated above, W. C. Allee's doctrine of social cooperation among animals was unconscious and involuntary; the process of cooperation was automatic. What appealed to Allee and to Ricketts was that this concept offered them an approach to reality that enabled them to break through to a view of the total picture. But seeing and understanding the whole picture, what Jim Casy calls "the whole shebang," and acting on the basis of that understanding, are two different things. *Sea of Cortez* enables us to see Ricketts and Steinbeck searching for and finding whole pictures. Steinbeck's novels and Ricketts's more recently published essays and articles provide us with a deeper understanding of the similarities and differences in their respective worldviews.

We read *Sea of Cortez* for its own sake as a first-rate work of travel literature. We read it also to understand the range and depth of Ricketts's impact on Steinbeck's fiction. And this permits us to see Steinbeck's fictional accomplishments in a new and fresh light. In so doing, we see not just the absurdity of arguments raised by those who attacked this or that Steinbeck novel on the basis of his alleged belief in any particular political ideology. We see also that his thinking is not worn and obsolete, but is as current as the modern environmental movement, which it predates and with which it has so much in common. If we read and consider *Sea of Cortez* in all its complexity, we see John Steinbeck fusing science and philosophy, art and ethics by combining the compelling if complex metaphysics of Ed Ricketts with his own commitment to social action by a species for whom he never gave up hope, and whom he believed could and would triumph over the tragic miracle of its own consciousness.

SUGGESTIONS FOR FURTHER READING

Allee, W. C. *Animal Aggregations*. Chicago: University of Chicago Press, 1931.

Astro, Richard. *John Steinbeck and Edward F. Ricketts: The Shaping of a Novelist*. Minneapolis: University of Minnesota Press, 1973.

———. *Edward F. Ricketts*. Western Writers Series. Boise, Ida.: Boise State University Press, 1976.

Benson, Jackson J. *The True Adventures of John Steinbeck*. New York: Viking Press, 1984.

Boodin, John Elof. *Cosmic Evolution*. New York: Macmillan Press, 1925.

Fadiman, Clifton. "Of Crabs and Men," *New Yorker*, December 6, 1941, 107.

Fontenrose, Joseph. *John Steinbeck: An Introduction and Interpretation*. New York: Barnes and Noble, 1964.

Hedgpeth, Joel W. "Philosophy on Cannery Row." In *Steinbeck: The Man and His Work*. Edited by Richard Astro and Tetsumaro Hayashi. Corvallis: Oregon State University Press, 1971.

Knox, Maxine, and Mary Rodriguez. *Steinbeck's Street: Cannery Row*. San Rafael, Calif.: Presidio Press, 1980.

Lisca, Peter. *The Wide World of John Steinbeck*. New Brunswick, N.J.: Rutgers University Press, 1958.

Lyman, John. "Of and About the Sea," *American Neptune*, April 1942, 183.

Mangelsdorf, Tom. *A History of Steinbeck's Cannery Row*. Santa Cruz, Calif.: Western Tanager Press, 1986.

Person, Richard. *History of Monterey*. Monterey, Calif.: City of Monterey, 1972.

Ricketts, Edward F., and Jack Calvin. *Between Pacific Tides*. 3d ed. Foreword by John Steinbeck. Stanford: Stanford University Press, 1952.

———. *The Outer Shores*. 2 vols. Edited by Joel W. Hedgpeth. Eureka, Calif.: Mad River Press, 1978.

———. "The Philosophy of Breaking Through." Unpublished MS, 1933.

———. "A Spiritual Morphology of Poetry." Unpublished MS, 1933.

———. "Thesis and Materials for a Script on Mexico." Unpublished MS, 1940.

Ritter, William Emerson. *The Unity of the Organism, or the Organismal Conception*. 2 vols. Boston: Gorham Press, 1919.

——— and Edna W. Bailey. *The Organismal Conception: Its Place in Science and Its Bearing on Philosophy*. Berkeley: University of California Publications in Zoology, 1931.

Steinbeck, Elaine, and Robert Walsten, eds. *Steinbeck: A Life in Letters*. New York: Viking Press, 1975.

Steinbeck, John. *Cannery Row*. New York: Viking Press, 1945.

———. *The Forgotten Village*. New York. Viking Press, 1941.

———. *The Grapes of Wrath*. New York: Viking Press, 1939.

A NOTE ON THE TEXT

The history of the publication of *Sea of Cortez* is interesting and chiefly involves the issue of joint authorship. Before the book was first published by Viking in December 1941, Steinbeck's editor, Pascal Covici, suggested that the title page read as follows:

The Sea of Cortez
By John Steinbeck
With a scientific appendix comprising materials for a source-book on the marine animals of the Panamic Faunal Province
By Edward F. Ricketts

Steinbeck objected vigorously, telling Covici that "this book is the product of the work and thinking of both of us and the setting down of the words is of no importance. . . . I not only disapprove of your plan—but forbid it."

The book was originally published as *Sea of Cortez: A Leisurely Journal of Travel and Research* by John Steinbeck and Edward F. Ricketts, with copyright in both authors' names. In 1951, the narrative portion of the book was published separately by Viking as *The Log from the Sea of Cortez*, with Steinbeck's preface "About Ed Ricketts." This Penguin Twentieth-Century Classics edition is based on the text of the 1951 publication; Steinbeck's "About Ed Ricketts" has been moved to the back matter as an appendix to the main text.

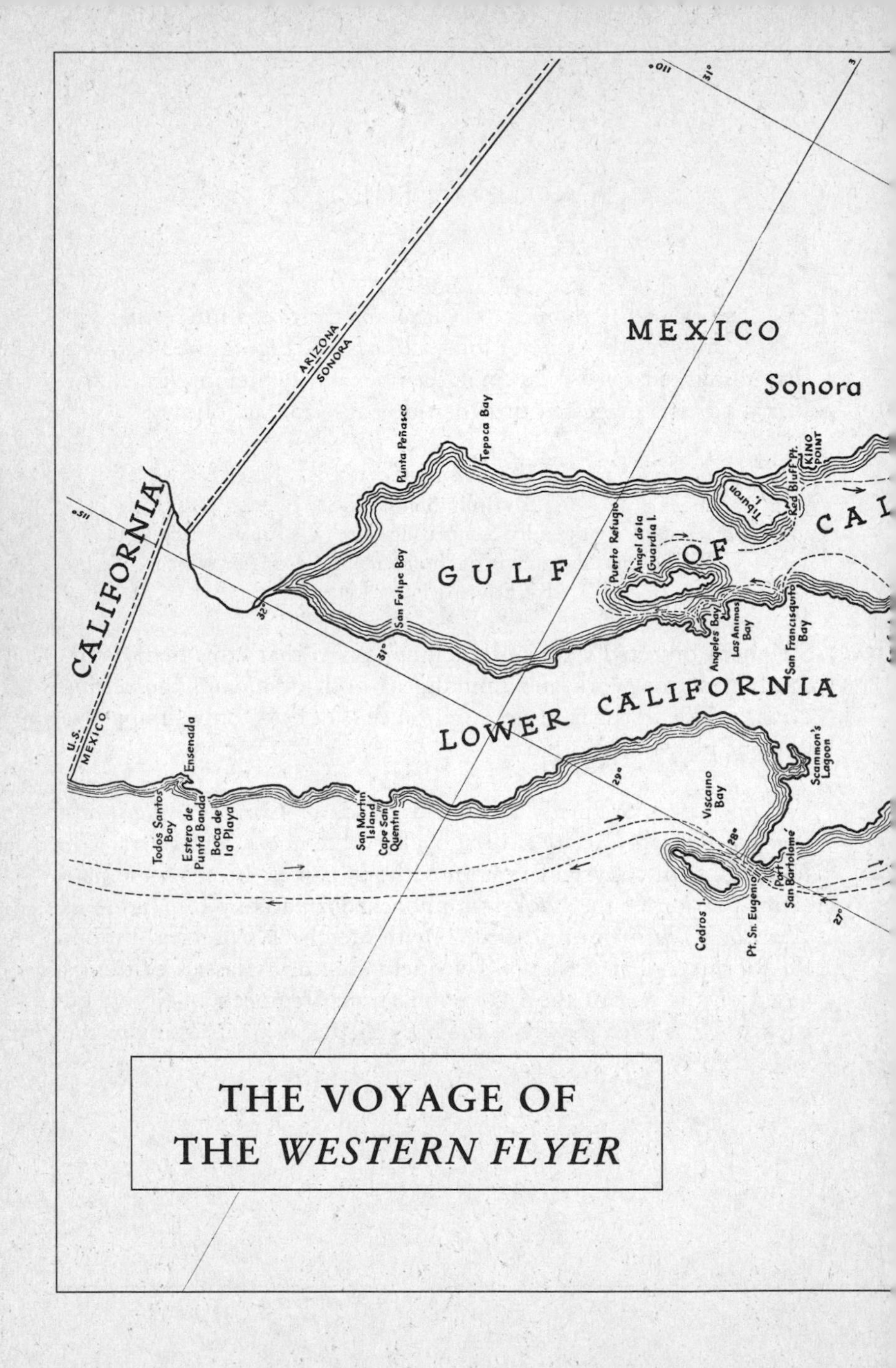

THE VOYAGE OF THE *WESTERN FLYER*

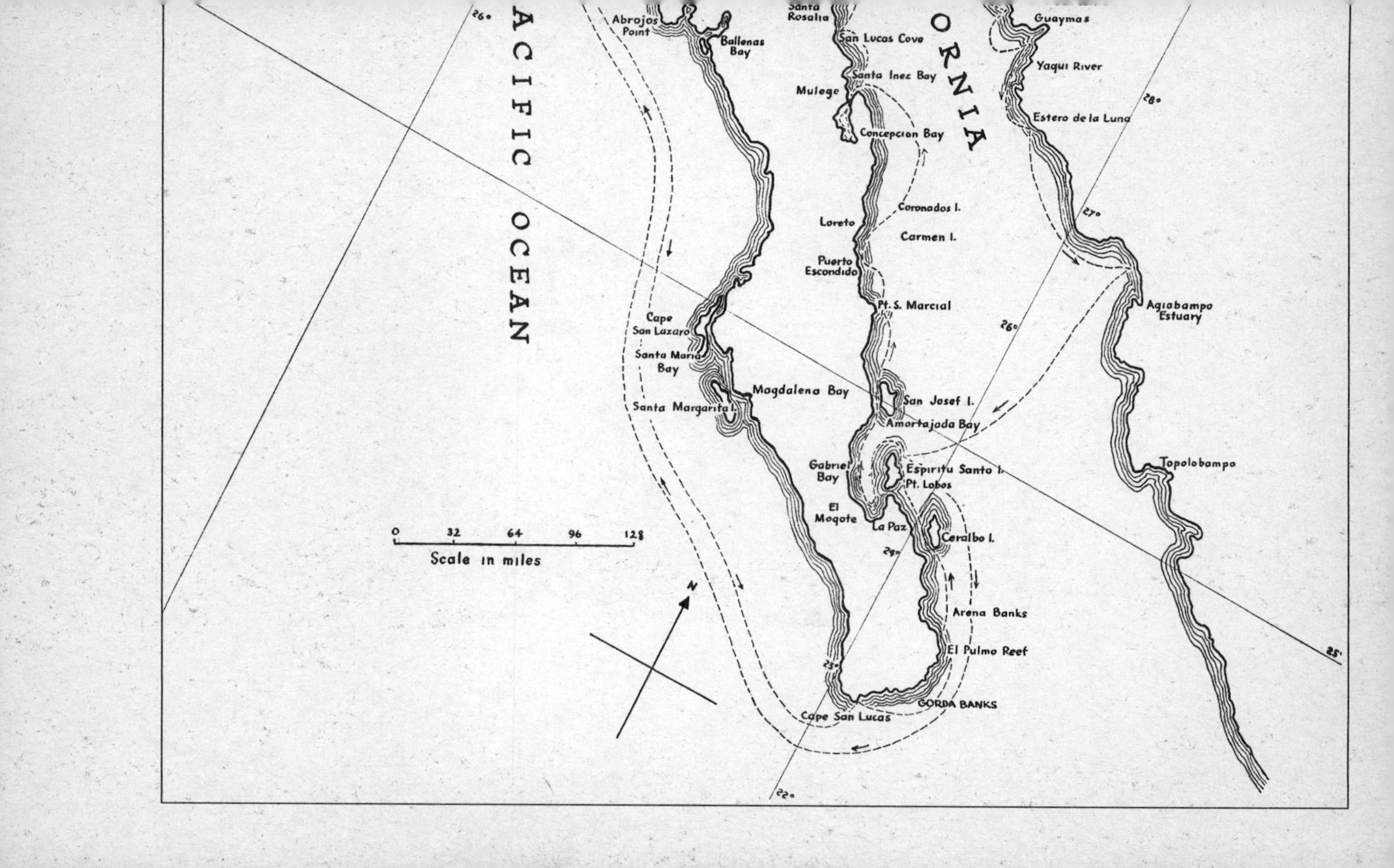
ORNIA
ACIFIC OCEAN
Abrojos Point
Ballenas Bay
Santa Rosalia
San Lucas Cove
Santa Inez Bay
Mulege
Concepcion Bay
Coronados I.
Carmen I.
Loreto
Puerto Escondido
Pt. S. Marcial
Cape San Lazaro
Santa Maria Bay
Magdalena Bay
Santa Margarita I.
San Josef I.
Amortajada Bay
Gabriel Bay
Espiritu Santo I.
Pt. Lobos
El Mogote
La Paz
Ceralbo I.
Arena Banks
El Pulmo Reef
GORDA BANKS
Cape San Lucas
Guaymas
Yaqui River
Estero de la Luna
Agiabampo Estuary
Topolobampo
0 32 64 96 128
Scale in miles
N

THE LOG FROM
THE *SEA OF CORTEZ*

INTRODUCTION

The design of a book is the pattern of a reality controlled and shaped by the mind of the writer. This is completely understood about poetry or fiction, but it is too seldom realized about books of fact. And yet the impulse which drives a man to poetry will send another man into the tide pools and force him to try to report what he finds there. Why is an expedition to Tibet undertaken, or a sea bottom dredged? Why do men, sitting at the microscope, examine the calcareous plates of a sea-cucumber, and, finding a new arrangement and number, feel an exaltation and give the new species a name, and write about it possessively? It would be good to know the impulse truly, not to be confused by the "services to science" platitudes or the other little mazes into which we entice our minds so that they will not know what we are doing.

We have a book to write about the Gulf of California. We could do one of several things about its design. But we have decided to let it form itself: its boundaries a boat and a sea; its duration a six weeks' charter time; its subject everything we could see and think and even imagine; its limits—our own without reservation.

We made a trip into the Gulf; sometimes we dignified it by calling it an expedition. Once it was called the Sea of Cortez, and that is a better-sounding and a more exciting name. We stopped in many little harbors and near barren coasts to collect and preserve the marine invertebrates of the littoral. One of the reasons we gave ourselves for this trip—and when we used this reason, we called the trip an expedition—was to observe the distribution of invertebrates, to see and to record their kinds and numbers, how they lived together, what they ate, and how they reproduced. That plan was simple, straight-forward, and only a part of the truth. But we did tell the truth to ourselves. We were curious. Our curiosity was not limited, but was as wide and horizonless as that of Darwin or Agassiz or Linnaeus or Pliny. We wanted to see everything our eyes would accommodate, to think what we could, and, out of our seeing and thinking, to build some kind of structure in modeled imitation of the observed reality. We knew that what we would see and record and construct would be warped, as all knowledge patterns are warped, first, by the collective pressure and stream of our time and race, second by the thrust of our individual personalities. But knowing this, we might not fall into too many holes—we might maintain some balance between our warp and the separate thing, the external reality. The oneness of these two might take its contribution from both. For example: the Mexican sierra has "XVII–15–IX" spines in the dorsal fin. These can easily be counted. But if the sierra strikes hard on the line so that our hands are burned, if the fish sounds and nearly escapes and finally comes in over the rail, his colors pulsing and his tail beating the air, a whole new relational externality has come into being—an entity which is more than the sum of the fish plus the fisherman. The only way to count the spines of the sierra unaffected by this second relational reality is to sit in a laboratory, open an evil-smelling jar, remove a stiff colorless fish from formalin solution, count the spines, and write the truth "D. XVII–15–IX." There you have recorded a reality which cannot be assailed—probably the least important reality concerning either the fish or yourself.

It is good to know what you are doing. The man with his pickled fish has set down one truth and has recorded in his experience many lies. The fish is not that color, that texture, that dead, nor does he smell that way.

Such things we had considered in the months of planning our expedition and we were determined not to let a passion for unassailable little truths draw in the horizons and crowd the sky down on us. We knew that what seemed to us true could be only relatively true anyway. There is no other kind of observation. The man with his pickled fish has sacrificed a great observation about himself, the fish, and the focal point, which is his thought on both the sierra and himself.

We suppose this was the mental provisioning of our expedition. We said, "Let's go wide open. Let's see what we see, record what we find, and not fool ourselves with conventional scientific strictures. We could not observe a completely objective Sea of Cortez anyway, for in that lonely and uninhabited Gulf our boat and ourselves would change it the moment we entered. By going there, we would bring a new factor to the Gulf. Let us consider that factor and not be betrayed by this myth of permanent objective reality. If it exists at all, it is only available in pickled tatters or in distorted flashes. Let us go," we said, "into the Sea of Cortez, realizing that we become forever a part of it; that our rubber boots slogging through a flat of eel-grass, that the rocks we turn over in a tide pool, make us truly and permanently a factor in the ecology of the region. We shall take something away from it, but we shall leave something too." And if we seem a small factor in a huge pattern, nevertheless it is of relative importance. We take a tiny colony of soft corals from a rock in a little water world. And that isn't terribly important to the tide pool. Fifty miles away the Japanese shrimp boats are dredging with overlapping scoops, bringing up tons of shrimps, rapidly destroying the species so that it may never come back, and with the species destroying the ecological balance of the whole region. That isn't very important in the world. And thousands of miles away the great bombs are falling and the stars are not moved thereby. None of it is important or all of it is.

We determined to go doubly open so that in the end we could, if we wished, describe the sierra thus: "D. XVII–15–IX; A. II–15–IX," but also we could see the fish alive and swimming, feel it plunge against the lines, drag it threshing over the rail, and even finally eat it. And there is no reason why either approach should be inaccurate. Spine-count description need not suffer because another approach is also used. Perhaps out of the two approaches, we thought, there might emerge a picture more complete and even more accurate than either alone could produce. And so we went.

1

How does one organize an expedition: what equipment is taken, what sources read; what are the little dangers and the large ones? No one has ever written this. The information is not available. The design is simple, as simple as the design of a well-written book. Your expedition will be enclosed in the physical framework of start, direction, ports of call, and return. These you can forecast with some accuracy; and in the better-known parts of the world it is possible to a degree to know what the weather will be in a given season, how high and low the tides, and the hours of their occurrence. One can know within reason what kind of boat to take, how much food will be necessary for a given crew for a given time, what medicines are usually needed—all this subject to accident, of course.

We had read what books were available about the Gulf and they were few and in many cases confused. The *Coast Pilot* had not been adequately corrected for some years. A few naturalists with specialties had gone into the Gulf and, in the way of specialists, had seen nothing they hadn't wanted to. Clavigero, a Jesuit of the eighteenth century, had seen more than most and reported what he saw with more accuracy than most. There were some romantic accounts by young people who had gone into the Gulf looking for adventure and, of course, had found it. The same romantic drive aimed at the stockyards would not be disappointed. From the information available, a few facts did emerge. The Sea of Cortez, or the Gulf of California, is a long, narrow, highly dangerous body of water. It is subject to sudden and vicious storms of great intensity. The months of March and April are usually quite calm and dependable and the March–April tides of 1940 were particularly good for collecting in the littoral.

The maps of the region were self-possessed and confident about headlands, coastlines, and depth, but at the edge of the Coast they become apologetic—laid in lagoons with dotted lines, supposed and presumed their boundaries. The *Coast Pilot* spoke as heatedly as it ever does about mirage and treachery of light. Going back from the *Coast Pilot* to Clavigero, we found more visual warnings in his accounts of ships broken up and scattered, of wrecks and wayward currents; of fifty miles of sea more dreaded than any other. The *Coast Pilot*, like an elderly scientist, cautious and restrained, on one side—and the old monk, setting down ships and men lost, and starvation on the inhospitable coasts.

In time of peace in the modern world, if one is thoughtful and careful, it is rather more difficult to be killed or maimed in the outland places of the globe than it is in the streets of our great cities, but the atavistic urge toward danger persists and its satisfaction is called adventure. However, your adventurer feels no gratification in crossing Market Street in San Francisco against the traffic. Instead he will go to a good deal of trouble and expense to get himself killed in the South Seas. In reputedly rough water, he will go in a canoe; he will invade deserts without adequate food and he will expose his tolerant and uninoculated blood to strange viruses. This is adventure. It is possible that his ancestor, wearying of the humdrum attacks of the saber-tooth, longed for the good old days of pterodactyl and triceratops.

We had no urge toward adventure. We planned to collect marine animals in a remote place on certain days and at certain hours indicated on the tide charts. To do this we had, in so far as we were able, to avoid adventure. Our plans, supplies, and equipment had to be more, not less, than adequate; and none of us was possessed of the curious boredom within ourselves which makes adventurers or bridge-players.

Our first problem was to charter a boat. It had to be sturdy and big enough to go to sea, comfortable enough to live on for six weeks, roomy enough to work on, and shallow enough so that little bays could be entered. The purse-seiners of Monterey were ideal for the purpose. They are dependable work boats with comfortable quarters and ample storage room. Furthermore, in March and April the sardine season is over and they are tied up. It would be easy, we thought, to charter such a boat; there must have been nearly a hundred of them anchored in back of the breakwater. We went to the pier and spread the word that we were looking for such a boat for charter. The word spread all right, but we were not overwhelmed with offers. In fact, no boat was offered. Only gradually did we discover the state of mind of the boat owners. They were uneasy about our project. Italians, Slavs, and some Japanese, they were primarily sardine fishers. They didn't even approve of fishermen who fished for other kinds of fish. They frankly didn't believe in the activities of the land—road-building and manufacturing and brick-laying. This was not a matter of ignorance on their part, but of intensity. All the directionalism of thought and emotion that man was capable of went into sardine-fishing; there wasn't room for anything else. An example of this occurred later when we were at sea. Hitler was invading Denmark and moving up towards Norway; there was no telling when the invasion of England might begin; our radio was full of static and the world was going to hell. Finally in all the crackle and noise of the short-wave one of our men made contact with another boat. The conversation went like this:

"This is the *Western Flyer*. Is that you, Johnny?"

"Yeah, that you, Sparky?"

"Yeah, this is Sparky. How much fish you got?"

"Only fifteen tons; we lost a school today. How much fish you got?"

"We're not fishing."

"Why not?"

"Aw, we're going down in the Gulf to collect starfish and bugs and stuff like that."

"Oh, yeah? Well, O.K., Sparky, I'll clear the wave length."

"Wait, Johnny. You say you only got fifteen tons?"

"That's right. If you talk to my cousin, tell him, will you?"

"Yeah, I will, Johnny. *Western Flyer's* all clear now."

Hitler marched into Denmark and into Norway, France had fallen, the Maginot Line was lost—we didn't know it, but we knew the daily catch of every boat within four hundred miles. It was simply a directional thing; a man has only so much. And so it was with the chartering of a boat. The owners were not distrustful of us; they didn't even listen to us because they couldn't quite believe we existed. We were obviously ridiculous.

Now the time was growing short and we began to worry. Finally one boat owner who was in financial difficulty offered his boat at a reasonable price and we were ready to accept when suddenly he raised the price out of question and bolted. He was horrified at what he had done. He raised the price, not to cheat us, but to get out of going.

The boat problem was growing serious when Anthony Berry sailed into Monterey Bay on the *Western Flyer*. The idea was no shock to Tony Berry; he had chartered to the government for salmon tagging in Alaskan waters and was used to nonsense. Besides, he was an intelligent and tolerant man. He knew that he had idiosyncrasies and that some of his friends had. He was willing to let us do any crazy thing that we wanted so long as we (1) paid a fair price, (2) told him where to go, (3) did not insist that he endanger the boat, (4) got back on time, and (5) didn't mix him up in our nonsense. His boat was not busy and he was willing to go. He was a quiet young man, very serious and a good master. He knew some navigation—a rare thing in the fishing fleet—and he had a natural caution which we admired. His boat was new and comfortable and clean, the engines in fine condition. We took the *Western Flyer* on charter.

She was seventy-six feet long with a twenty-five-foot beam; her engine, a hundred and sixty-five horsepower direct reversible Diesel, drove her at ten knots. Her deckhouse had a wheel forward, then combination master's room and radio room, then bunkroom, very comfortable, and behind that the galley. After the galley, a large hatch gave into the fish-hold, and after the hatch were the big turn-table and roller of the purse-seiner. She carried a twenty-foot skiff and a ten-foot skiff. Her engine was a thing of joy, spotlessly clean, the moving surfaces shining and damp with oil and the green paint fresh and new on the housings. The engine-room floor was clean and all the tools polished and hung in their places. One look into the engine-room inspired confidence in the master. We had seen other engines in the fishing fleet and this perfection on the *Western Flyer* was by no means a general thing.

As crew we signed Tex Travis, engineer, and Sparky Enea and Tiny Colletto, seamen. All three were a little reluctant to go, for the whole thing was crazy. None of us had been into the Gulf, although the master had been as far as Cape San Lucas, and the Gulf has a really bad name. It was a thoughtful crew who agreed to go with us.

We could never tell when the change of attitude toward us came, but it came very rapidly. Perhaps it was because Tony Berry was known as a cautious man who would not indulge in nonsense, or perhaps it was pure relief that at last it had been settled. All of a sudden we were overwhelmed with help. We had offers from men to go with us without pay. Sparky was offered a certain price for his job that was more than he would get from us. All he had to do was turn over his job and sit in Monterey and spend the money. But Sparky refused. Our project had become honorable. We had more help than we could use and advice enough to move the navies of the world.

We did not know what our crew thought of the expedition but later, in the field, they became good collectors—a little emotional sometimes, as when Tiny, in outrage at being pinched, declared a war of extermination on the whole Sally Lightfoot species, but on the whole collectors of taste and quickness.

The charter was signed with dignity and reverence. It is impossible to be light-hearted in the face of a ship's charter, for the law has foreseen or remembered the most doleful and arbitrary acts of God and has set them down as possibilities, but in the tone of inevitabilities. Thus, you read what you or the others must do in the case of wreck, or sunken rocks; of death at sea in its most painful and astonishing aspects; of injury to plank and keel; of water shortage and mutiny. Next to marriage settlement or sentence of death, a ship's charter is as portentous a document as has ever been written. Penalties are set down against both parties, and if on some morning the rising sun should find your ship in the middle of the Mojave Desert you have only to look again at the charter to find the blame assigned and the penalty indicated. It took us several hours to get over the solemn feeling the charter put on us. We thought we might live better lives and pay our debts, and one at least of us contemplated for one holy, horrified moment a vow of chastity.

But the charter was signed and food began to move into the *Western Flyer*. It is amazing how much food seven people need to exist for six weeks. Cases of spaghetti, cases and cases of peaches and pineapple, of tomatoes, whole Romano cheeses, canned milk in coveys, flour and cornmeal, gallons of olive oil, tomato paste, crackers, cans of butter and jam, catsup and rice, beans and bacon and canned meats, vegetables and soups in cans; truckloads of food. And all this food was stored eagerly and happily by the crew. It disappeared into cupboards, under little hatches in the galley floor, and many cases went below.

We had done a good deal of collecting, but largely in temperate zones. The equipment for collecting, preserving, and storing specimens was selected on the basis of experience in other waters and of anticipation of difficulties imposed by a hot humid country. In some cases we were right, in others very wrong.

In a small boat, the library should be compact and available. We had constructed a strong, steel-reinforced wooden case, the front of which hinged down to form a desk. This case holds about twenty large volumes and has two filing cases, one for separates (scientific reprints) and one for letters; a small metal box holds pens, pencils, erasers, clips, steel tape, scissors, labels, pins, rubber bands, and so forth. Another compartment contains a three-by-five-inch card file. There are cubby-holes for envelopes, large separates, small separates, typewriter paper, carbon, a box for India ink and glue. The construction of the front makes room for a portable typewriter, drawing board, and T-square. There is a long narrow space for rolled charts and maps. Closed, this compact and complete box is forty-four inches long by eighteen by eighteen; loaded, it weighs between three and four hundred pounds. It was designed to rest on a low table or in an unused bunk. Its main value is compactness, completeness, and accessibility. We took it aboard the *Western Flyer*. There was no table for it to rest on. It did not fit in a bunk. It could not be put on the deck because of moisture. It ended up lashed to the rail on top of the deckhouse, covered with several layers of tarpaulin and roped on. Because of the roll of the boat it had to be tied down at all times. It took about ten minutes to remove the tarpaulin, untie the lashing line, open the cover, squeeze down between two crates of oranges, read the title of the wanted book upside down, remove it, close and lash and cover the box again. But if there had been a low table or a large bunk, it would have been perfect.

For many little errors like this, we have concluded that all collecting trips to fairly unknown regions should be made twice; once to make mistakes and once to correct them. Some of the greatest difficulty lies in the fact that previous collectors have never set down the equipment taken and its success or failure. We propose to rectify this in our account.

The library contained all the separates then available on the Panamic and Gulf fauna. Primary volumes such as Johnson and Snook, Ricketts and Calvin, Russell and Yonge, Flattely and Walton, Keep's *West Coast Shells*, Fisher's three-volume starfish monograph, the Rathbun brachyuran monograph, Schmitt's *Marine Decapod Crustacea of California*, Fraser's *Hydroids*, Barnhart's *Marine Fishes of Southern California*, *Coast Pilots* for the whole Pacific Coast; charts, both large and small scale, of the whole region to be covered.

The camera equipment was more than adequate, for it was never used. It included a fine German reflex and an 8-mm. movie camera with tripod, light meters, and everything. But we had no camera-man. During low tides we all collected; there was no time to dry hands and photograph at the collecting scene. Later, the anesthetizing, killing, preserving, and labeling of specimens were so important that we still took no pictures. It was an error in personnel. There should be a camera-man who does nothing but take pictures.

Our collecting material at least was good. Shovels, wrecking- and abalone-bars, nets, long-handled dip-nets, wooden fish-kits, and a number of seven-cell flashlights for night collecting were taken. Containers seemed to go endlessly into the hold of the *Western Flyer*. Wooden fish-kits with heads; twenty hard-fir barrels with galvanized hoops in fifteen- and thirty-gallon sizes; cases of gallon jars, quart, pint, eight-ounce, five-ounce, and two-ounce screw-cap jars; several gross of corked vials in four chief sizes, 100×33 mm., six-dram, four-dram, and two-dram sizes. There were eight two-and-a-half-gallon jars with screw caps. And with all these we ran short of containers, and before we were through had to crowd those we had. This was unfortunate, since many delicate animals should be preserved separately to prevent injury.

Of chemicals, we put into the boat a fifteen-gallon barrel of U.S.P. formaldehyde and a fifteen-gallon barrel of denatured alcohol. This was not nearly enough alcohol. The stock had to be replenished at Guaymas, where we bought ten gallons of pure sugar alcohol. We took two gallons of Epsom salts for anesthetization and again ran out and had to buy more in Guaymas. Menthol, chromic acid, and novocain, all for relaxing animals, were included in the chemical kit. Of preparing equipment, there were glass chiton plates and string, lots of rubber gloves, graduates, forceps, and scalpels. Our binocular microscope, Bausch & Lomb A.K.W., was fitted with a twelve-volt light, but on the rolling boat the light was so difficult to handle that we used a spot flashlight instead. We had galvanized iron nested trays of fifteen- to twenty-gallon capacity for gross hardening and preservation. We had enameled and glass trays for the laying out of specimens, and one small examination aquarium.

The medical kit had been given a good deal of thought. There were nembutal, butesin picrate for sunburn, a thousand two-grain quinine capsules, two-percent mercuric oxide salve for barnacle cuts, cathartics, ammonia, mercurochrome, iodine, alcaroid, and, last, some whisky for medicinal purposes. This did not survive our leave-taking, but since no one was ill on the whole trip, it may have done its job very well.

2

What little time we were not on lists and equipment or in grudging sleep we went to the pier and looked at boats, watched them tied to their buoys behind the breakwater—the dirty boats and the clean painted boats, each one stamped with the personality of its owner. Here, where the discipline was as individual as the owners, every boat was different from every other one. If the stays were rusting and the deck unwashed, paint scraped off and lines piled carelessly, there was no need to see the master; we knew him. And if the lines were coiled and the cables greased and the little luxury of deer horns nailed to the crow's-nest, there was no need to see that owner either. There were deer horns on many of the crow's-nests, and when we asked why, we were told they brought good luck. Out of some ancient time, they brought good luck to these people, most of them out of Sicily, the horns grown sturdily on the structure of their race. If you ask, "Where does the idea come from?" the owner will say, "It brings good luck, we always put them on." And a thousand years ago the horns were on the masts and brought good luck, and probably when the ships of Carthage and Tyre put into the harbors of Sicily, the horns were on the mastheads and brought good luck and no one knew why. Out of some essential race soul the horns come, and not only the horns, but the boats themselves, so that to a man, to nearly all men, a boat more than any other tool he uses is a little representation of an archetype. There is an "idea" boat that is an emotion, and because the emotion is so strong it is probable that no other tool is made with so much honesty as a boat. Bad boats are built, surely, but not many of them. It can be argued that a bad boat cannot survive tide and wave and hence is not worth building, but the same might be said of a bad automobile on a rough road. Apparently the builder of a boat acts under a compulsion greater than himself. Ribs are strong by definition and feeling. Keels are sound, planking truly chosen and set. A man builds the best of himself into a boat—builds many of the unconscious memories of his ancestors. Once, passing the boat department of Macy's in New York, where there are duck-boats and skiffs and little cruisers, one of the authors discovered that as he passed each hull he knocked on it sharply with his knuckles. He wondered why he did it, and as he wondered, he heard a knocking behind him, and another man was rapping the hulls with *his* knuckles, the same tempo—three sharp knocks on each hull. During an hour's observation there no man or boy and few women passed who did not do the same thing. Can this have been an unconscious testing of the hulls? Many who passed could not have been in a boat, perhaps some of the little boys had never seen a boat, and yet everyone tested the hulls, knocked to see if they were sound, and did not even know he was doing it. The observer thought perhaps they and he would knock on any large wooden object that might give forth a resonant sound. He went to the piano department, icebox floor, beds, cedar-chests, and no one knocked on them—only on boats.

How deep this thing must be, the giver and the receiver again; the boat designed through millenniums of trial and error by the human consciousness, the boat which has no counterpart in nature unless it be a dry leaf fallen by accident in a stream. And Man receiving back from Boat a warping of his psyche so that the sight of a boat riding in the water clenches a fist of emotion in his chest. A horse, a beautiful dog, arouses sometimes a quick emotion, but of inanimate things only a boat can do it. And a boat, above all other inanimate things, is personified in man's mind. When we have been steering, the boat has seemed sometimes nervous and irritable, swinging off course before the correction could be made, slapping her nose into the quartering wave. After a storm she has seemed tired and sluggish. Then with the colored streamers set high and snapping, she is very happy, her nose held high and her stern bouncing a little like the buttocks of a proud and confident girl. Some have said they have felt a boat shudder before she struck a rock, or cry when she beached and the surf poured into her. This is not mysticism, but identification; man, building this greatest and most personal of all tools, has in turn received a boat-shaped mind, and the boat, a man-shaped soul. His spirit and the tendrils of his feeling are so deep in a boat that the identification is complete. It is very easy to see why the Viking wished his body to sail away in an unmanned ship, for neither could exist without the other; or, failing that, how it was necessary that the things he loved most, his women and his ship, lie with him and thus keep closed the circle. In the great fire on the shore, all three started at least in the same direction, and in the gathered ashes who could say where man or woman stopped and ship began?

This strange identification of man with boat is so complete that probably no man has even destroyed a boat by bomb or torpedo or shell without murder in his heart; and were it not for the sad trait of self-destruction that is in our species, he could not do it. Only the trait of murder which our species seems to have could allow us the sick, exultant sadness of sinking a ship, for we can murder the things we love best, which are, of course, ourselves.

We have looked into the tide pools and seen the little animals feeding and reproducing and killing for food. We name them and describe them and, out of long watching, arrive at some conclusion about their habits so that we say, "This species typically does thus and so," but we do not objectively observe our own species as a species, although we know the individuals fairly well. When it seems that men may be kinder to men, that wars may not come again, we completely ignore the record of our species. If we used the same smug observation on ourselves that we do on hermit crabs we would be forced to say, with the information at hand, "It is one diagnostic trait of *Homo sapiens* that groups of individuals are periodically infected with a feverish nervousness which causes the individual to turn on and destroy, not only his own kind, but the works of his own kind. It is not known whether this be caused by a virus, some airborne spore, or whether it be a species reaction to some meteorological stimulus as yet undetermined." Hope, which is another species diagnostic trait—the hope that this may not always be—does not in the least change the observable past and present. When two crayfish meet, they usually fight. One would say that perhaps they might not at a future time, but without some mutation it is not likely that they will lose this trait. And perhaps our species is not likely to forgo war without some psychic mutation which at present, at least, does not seem imminent. And if one place the blame for killing and destroying on economic insecurity, on inequality, on injustice, he is simply stating the proposition in another way. We have what we are. Perhaps the crayfish feels the itch of jealousy, or perhaps he is sexually insecure. The effect is that he fights. When in the world there shall come twenty, thirty, fifty years without evidence of our murder trait, under whatever system of justice or economic security, then we may have a contrasting habit pattern to examine. So far there is no such situation. So far the murder trait of our species is as regular and observable as our various sexual habits.

3

In the time before our departure for the Gulf we sat on the pier and watched the sardine purse-seiners riding among the floating grapefruit rinds. A breakwater is usually a dirty place, as though the tampering with the shore line is obscene and impractical to the cleansing action of the sea. And we talked to our prospective crew. Tex, our engineer, was caught in the ways of the harbor. He was born in the Panhandle of Texas and early he grew to love Diesel engines. They are so simple and powerful, blocks of pure logic in shining metal. They appealed to some sense of neat thinking in Tex. He might be sentimental and illogical in some things, but he liked his engines to be true and logical. By an accident, possibly alcoholic, he came to the Coast in an old Ford and sat down beside the Bay, and there he discovered a wonderful thing. Here, combined in one, were the best Diesels to be found anywhere, and boats. He never recovered from his shocked pleasure. He could never leave the sea again, for nowhere else could he find these two perfect things in one. He is a sure man with an engine. When he goes below he is identified with his engine. He moves about, not seeing, not looking, but knowing. No matter how tired or how deeply asleep he may be, one miss of the engine jerks him to his feet and into the engine-room before he is awake, and we truly believe that a burned bearing or a cracked shaft gives him sharp pains in his stomach.

We talked to Tony, the master and part owner of the *Western Flyer*, and our satisfaction with him as master increased constantly. He had the brooding, dark, Slavic eyes and the hawk nose of the Dalmatian. He rarely talked or laughed. He was tall and lean and very strong. He had a great contempt for forms. Under way, he liked to wear a tweed coat and an old felt hat, as though to say, "I keep the sea in my head, not on my back like a Goddamn yachtsman." Tony has one great passion; he loves rightness and he hates wrongness. He thinks speculation a complete waste of time. To our sorrow, and some financial loss, we discovered that Tony never spoke unless he was right. It was useless to bet with him and impossible to argue with him. If he had not been right, he would never have opened his mouth. But once knowing and saying a truth, he became infuriated at the untruth which naturally enough was set against it. Inaccuracy was like an outrageous injustice to him, and when confronted with it, he was likely to shout and to lose his temper. But he did not personally triumph when his point was proven. An ideal judge, hating larceny, feels no triumph when he sentences a thief, and Tony, when he has nailed a true thing down and routed a wrong thing, feels good, but not righteous. He retires grumbling a little sadly at the stupidity of a world which can conceive a wrongness or for one moment defend one. He loves the leadline because it tells a truth on its markers; he loves the Navy charts; and until he went into the Gulf he admired the *Coast Pilot*. The *Coast Pilot* was not wrong, but things had changed since its correction, and Tony is uneasy in the face of variables. The whole relational thinking of modern physics was an obscenity to him and he refused to have anything to do with it. Parallels and compasses and the good Navy maps were things you could trust. A circle is true and a direction is set forever, a shining golden line across the mind. Later, in the mirage of the Gulf where visual distance is a highly variable matter, we wondered whether Tony's certainties were ever tipped. It did not seem so. His qualities made him a good master. He took no chances he could avoid, for his boat and his life and ours were no light things for him to tamper with.

We come now to a piece of equipment which still brings anger to our hearts and, we hope, some venom to our pen. Perhaps in self-defense against suit, we should say, "The outboard motor mentioned in this book is purely fictitious and any resemblance to outboard motors living or dead is coincidental." We shall call this contraption, for the sake of secrecy, a Hansen Sea-Cow—a dazzling little piece of machinery, all aluminum paint and touched here and there with spots of red. The Sea-Cow was built to sell, to dazzle the eyes, to splutter its way into the unwary heart. We took it along for the skiff. It was intended that it should push us ashore and back, should drive our boat into estuaries and along the borders of little coves. But we had not reckoned with one thing. Recently, industrial civilization has reached its peak of reality and has lunged forward into something that approaches mysticism. In the Sea-Cow factory where steel fingers tighten screws, bend and mold, measure and divide, some curious mathematick has occurred. And that secret so long sought has accidentally been found. Life has been created. The machine is at last stirred. A soul and a malignant mind have been born. Our Hansen Sea-Cow was not only a living thing but a mean, irritable, contemptible, vengeful, mischievous, hateful living thing. In the six weeks of our association we observed it, at first mechanically and then, as its living reactions became more and more apparent, psychologically. And we determined one thing to our satisfaction. When and if these ghoulish little motors learn to reproduce themselves the human species is doomed. For their hatred of us is so great that they will wait and plan and organize and one night, in a roar of little exhausts, they will wipe us out. We do not think that Mr. Hansen, inventor of the Sea-Cow, father of the outboard motor, knew what he was doing. We think the monster he created was as accidental and arbitrary as the beginning of any other life. Only one thing differentiates the Sea-Cow from the life that we know. Whereas the forms that are familiar to us are the results of billions of years of mutation and complication, life and intelligence emerged simultaneously in the Sea-Cow. It is more than a species. It is a whole new redefinition of life. We observed the following traits in it and we were able to check them again and again:

1. Incredibly lazy, the Sea-Cow loved to ride on the back of a boat, trailing its propeller daintily in the water while we rowed.

2. It required the same amount of gasoline whether it ran or not, apparently being able to absorb this fluid through its body walls without recourse to explosion. It had always to be filled at the beginning of every trip.

3. It had apparently some clairvoyant powers, and was able to read our minds, particularly when they were inflamed with emotion. Thus, on every occasion when we were driven to the point of destroying it, it started and ran with a great noise and excitement. This served the double purpose of saving its life and of resurrecting in our minds a false confidence in it.

4. It had many cleavage points, and when attacked with a screwdriver, fell apart in simulated death, a trait it had in common with opossums, armadillos, and several members of the sloth family, which also fall apart in simulated death when attacked with a screwdriver.

5. It hated Tex, sensing perhaps that his knowledge of mechanics was capable of diagnosing its shortcomings.

6. It completely refused to run: (a) when the waves were high, (b) when the wind blew, (c) at night, early morning, and evening, (d) in rain, dew, or fog, (e) when the distance to be covered was more than two hundred yards. But on warm, sunny days when the weather was calm and the white beach close by—in a word, on days when it would have been a pleasure to row—the Sea-Cow started at a touch and would not stop.

7. It loved no one, trusted no one. It had no friends.

Perhaps toward the end, our observations were a little warped by emotion. Time and again as it sat on the stern with its pretty little propeller lying idly in the water, it was very close to death. And in the end, even we were infected with its malignancy and its dishonesty. We should have destroyed it, but we did not. Arriving home, we gave it a new coat of aluminum paint, spotted it at points with new red enamel, and sold it. And we might have rid the world of this mechanical cancer!

4

It would be ridiculous to suggest that ours was anything but a makeshift expedition. The owner of a boat on short charter does not look happily on any re-designing of his ship. In a month or two we could have changed the *Western Flyer* about and made her a collector's dream, but we had neither the time nor the money to do it. The low-tide period was approaching. We had on board no permanent laboratory. There was plenty of room for one in the fish-hold, but the dampness there would have rusted the instruments overnight. We had no dark-room, no permanent aquaria, no tanks for keeping animals alive, no pumps for delivering sea water. We had not even a desk except the galley table. Microscopes and cameras were put away in an empty bunk. The enameled pans for laying out animals were in a large crate lashed to the net-table aft, where it shared the space with the two skiffs. The hatch cover of the fish-hold became laboratory and aquarium, and we carried sea water in buckets to fill the pans. Another empty bunk was filled with flashlights, medicines, and the more precious chemicals. Dip-nets, wooden collecting buckets, and vials and jars in their cases were stowed in the fish-hold. The barrels of alcohol and formaldehyde were lashed firmly to the rail on deck, for all of us had, I think, a horror-thought of fifteen gallons of U.S.P. formaldehyde broken loose and burst. One achieves a respect and a distaste for formaldehyde from working with it. Fortunately, none of us had a developed formalin allergy. Our small refrigerating chamber, powered by a two-cycle gasoline engine and designed to cool sea water for circulation to living animals, began the trip on top of the deck-house and ended back on the net-table. This unit, by the way, was not very effective, the motor being jerky and not of sufficient power. But on certain days in the Gulf it did manage to cool a little beer or perhaps more than a little, for the crew fell in joyfully with our theory that it is unwise to drink unboiled water, and boiled water isn't any good. In addition, the weather was too hot to boil water, and besides the crew wished to test this perfectly sound scientific observation thoroughly. We tested it by reducing the drinking of water to an absolute minimum.

A big pressure tube of oxygen was lashed to a deck rail, its gauges and valves wrapped in canvas. Gradually, the boat was loaded and the materials put away, some never to be taken out again. It was agreed that we should all stand wheel-watch when we were running night and day; but once in the Gulf, and working at collecting stations, the hired crew should work the boat, since we would anchor at night and run only during the daytime.

Toward the end of the preparation, a small hysteria began to build in ourselves and our friends. There were hundreds of unnecessary trips back and forth. Some materials were stowed on board with such cleverness that we never found them again. Now the whole town of Monterey was becoming fevered and festive—but not because of our going. At the end of the sardine season, canneries and boat owners provide a celebration. There is a huge barbecue on the end of the pier with free beef and beer and salad for all comers. The sardine fleet is decorated with streamers and bunting and serpentine, and the boat with the biggest season catch is queen of a strange nautical parade of boats; and every boat is an open house, receiving friends of owners and of crew. Wine flows beautifully, and the parade of boats that starts with dignity and precision sometimes ends in a turmoil. This fiesta took place on Sunday, and we were to sail on Monday morning. The *Western Flyer* was decorated like the rest with red and blue bunting and serpentine. Master and crew refused to sail before the fiesta was over. We rode in the parade of boats, some of us in the crow's-nest and some on the house. With five thousand other people we crowded on the pier and ate great hunks of meat and drank beer and heard speeches. It was the biggest barbecue the sardine men had ever given, and the potato salad was served out of washtubs. The speeches rose to a crescendo of patriotism and good feeling beyond anything Monterey had ever heard.

There should be here some mention of the permits obtained from the Mexican government. At the time of our preparation, Mexico was getting ready for a presidential election, and the apparent issues were so complex as to cause apprehension that there might be violence. The nation was a little nervous, and it seemed to us that we should be armed with permits which clearly established us as men without politics or business interests. The work we intended to do might well have seemed suspicious to some patriotic customs official or soldier—a small boat that crept to uninhabited points on a barren coast, and a party which spent its time turning over rocks. It was not likely that we could explain our job to the satisfaction of a soldier. It would seem ridiculous to the military mind to travel fifteen hundred miles for the purpose of turning over rocks on the seashore and picking up small animals, very few of which were edible; and doing all this without shooting at anyone. Besides, our equipment might have looked subversive to one who had seen the war sections of *Life* and *Pic* and *Look*. We carried no firearms except a .22-caliber pistol and a very rusty ten-gauge shotgun. But an oxygen cylinder might look too much like a torpedo to an excitable rural soldier, and some of the laboratory equipment could have had a lethal look about it. We were not afraid for ourselves, but we imagined being held in some mud *cuartel* while the good low tides went on and we missed them. In our naïveté, we considered that our State Department, having much business with the Mexican government, might include a paragraph about us in one of its letters, which would convince Mexico of our decent intentions. To this end, we wrote to the State Department explaining our project and giving a list of people who would confirm the purity of our motives. Then we waited with a childlike faith that when a thing is stated simply and evidence of its truth is included there need be no mix-up. Besides, we told ourselves, we were American citizens and the government was our servant. Alas, we did not know diplomatic procedure. In due course, we had an answer from the State Department. In language so diplomatic as to be barely intelligible it gently disabused us. In the first place, the State Department was *not* our servant, however other departments might feel about it. The State Department had little or no interest in the collection of marine invertebrates unless carried on by an institution of learning, preferably with Dr. Butler as its president. The government never made such representations for private citizens. Lastly, the State Department hoped to God we would not get into trouble and appeal to it for aid. All this was concealed in language so beautiful and incomprehensible that we began to understand why diplomats say they are "studying" a message from Japan or England or Italy. We studied this letter for the better part of one night, reduced its sentences to words, built it up again, and came out with the above-mentioned gist. "Gist" is, we imagine, a word which makes the State Department shudder with its vulgarity.

There we were, with no permits and the imaginary soldier still upset by our oxygen tube. In Mexico, certain good friends worked to get us the permits; the consul-general in San Francisco wrote letters about us, and then finally, through a friend, we got in touch with Mr. Castillo Najera, the Mexican ambassador to Washington. To our wonder there came an immediate reply from the ambassador which said there was no reason why we should not go and that he would see the permits were issued immediately. His letter said just that. There was a little sadness in us when we read it. The ambassador seemed such a good man we felt it a pity that he had no diplomatic future, that he could never get anywhere in the world of international politics. We understood his letter the first time we read it. Clearly, Mr. Castillo Najera is a misfit and a rebel. He not only wrote clearly, but he kept his word. The permits came through quickly and in order. And we wish here and now to assure this gentleman that whenever the inevitable punishment for his logic and clarity falls upon him we will gladly help him to get a new start in some other profession.

When the permits arrived, they were beautifully sealed so that even a soldier who could not read would know that if we were not what we said we were, we were at least influential enough spies and saboteurs to be out of his jurisdiction.

And so our boat was loaded, except for the fuel tanks, which we planned to fill at San Diego. Our crew entered the contests at the sardine fiesta—the skiff race, the greased-pole walk, the water-barrel tilt—and they did not win anything, but no one cared. And late in the night when the feast had died out we slept ashore for the last time, and our dreams were cluttered with things we might have forgotten. And the beer cans from the fiesta washed up and down the shore on the little brushing waves behind the breakwater.

We had planned to sail about ten o'clock on March 11, but so many people came to see us off and the leave-taking was so pleasant that it was afternoon before we could think of going. The moment or hour of leave-taking is one of the pleasantest times in human experience, for it has in it a warm sadness without loss. People who don't ordinarily like you very well are overcome with affection at leave-taking. We said good-by again and again and still could not bring ourselves to cast off the lines and start the engines. It would be good to live in a perpetual state of leave-taking, never to go nor to stay, but to remain suspended in that golden emotion of love and longing; to be missed without being gone; to be loved without satiety. How beautiful one is and how desirable; for in a few moments one will have ceased to exist. Wives and fiancées were there, melting and open. How beautiful they were too; and against the hull of the boat the beer cans from the fiesta of yesterday tapped lightly like little bells, and the sea-gulls flew around and around but did not land. There was no room for them—too many people were seeing us off. Even a few strangers were caught in the magic and came aboard and wrung our hands and went into the galley. If our medicine chest had held out we might truly never have sailed. But about twelve-thirty the last dose was prescribed and poured and taken. Only then did we realize that not only were *we* fortified against illness, but that fifty or sixty inhabitants of Monterey could look forward to a long period of good health.

The day of charter had arrived. That instrument said we would leave on the eleventh, and the master was an honest man. We ejected our guests, some forcibly. The lines were cast off. We backed and turned and wove our way out among the boats of the fishing fleet. In our rigging the streamers, the bunting, the serpentine still fluttered, and as the breakwater was cleared and the wind struck us, we seemed, to ourselves at least, a very brave and beautiful sight. The little bell buoy on the reef at Cabrillo Point was excited about it too, for the wind had freshened and the float rolled heavily and the four clappers struck the bell with a quick tempo. We stood on top of the deckhouse and watched the town of Pacific Grove slip by and dark pine-covered hills roll back on themselves as though they moved, not we.

We sat on a crate of oranges and thought what good men most biologists are, the tenors of the scientific world—temperamental, moody, lecherous, loud-laughing, and healthy. Once in a while one comes on the other kind—what used in the university to be called a "dry-ball"—but such men are not really biologists. They are the embalmers of the field, the picklers who see only the preserved form of life without any of its principle. Out of their own crusted minds they create a world wrinkled with formaldehyde. The true biologist deals with life, with teeming boisterous life, and learns something from it, learns that the first rule of life is living. The dry-balls cannot possibly learn a thing every starfish knows in the core of his soul and in the vesicles between his rays. He must, so know the starfish and the student biologist who sits at the feet of living things, proliferate in all directions. Having certain tendencies, he must move along their lines to the limit of their potentialities. And we have known biologists who did proliferate in all directions: one or two have had a little trouble about it. Your true biologist will sing you a song as loud and off-key as will a blacksmith, for he knows that morals are too often diagnostic of prostatitis and stomach ulcers. Sometimes he may proliferate a little too much in all directions, but he is as easy to kill as any other organism, and meanwhile he is very good company, and at least he does not confuse a low hormone productivity with moral ethics.

The *Western Flyer* pushed through the swells toward Point Joe, which is the southern tip of the Bay of Monterey. There was a line of white which marked the open sea, for a strong north wind was blowing, and on that reef the whistling buoy rode, roaring like a perplexed and mournful bull. On the shore road we could see the cars of our recent friends driving along keeping pace with us while they waved handkerchiefs sentimentally. We were all a little sentimental that day. We turned the buoy and cleared the reef, and as we did the boat rolled heavily and then straightened. The north wind drove down on our tail, and we headed south with the big swells growing under us and passing, so that we seemed to be standing still. A squadron of pelicans crossed our bow, flying low to the waves and acting like a train of pelicans tied together, activated by one nervous system. For they flapped their powerful wings in unison, coasted in unison. It seemed that they tipped a wavetop with their wings now and then, and certainly they flew in the troughs of the waves to save themselves from the wind. They did not look around or change direction. Pelicans seem always to know exactly where they are going. A curious sea-lion came out to look us over, a tawny, crusty old fellow with rakish mustaches and the scars of battle on his shoulders. He crossed our bow too and turned and paralleled our course, trod water, and looked at us. Then, satisfied, he snorted and cut for shore and some sea-lion appointment. They always have them, it's just a matter of getting around to keeping them.

And now the wind grew stronger and the windows of houses along the shore flashed in the declining sun. The forward guy-wire of our mast began to sing under the wind, a deep and yet penetrating tone like the lowest string of an incredible bull-fiddle. We rose on each swell and skidded on it until it passed and dropped us in the trough. And from the galley ventilator came the odor of boiling coffee, a smell that never left the boat again while we were on it.

In the evening we came back restlessly to the top of the deckhouse, and we discussed the Old Man of the Sea, who might well be a myth, except that too many people have seen him. There is some quality in man which makes him people the ocean with monsters and one wonders whether they are there or not. In one sense they are, for we continue to see them. One afternoon in the laboratory ashore we sat drinking coffee and talking with Jimmy Costello, who is a reporter on the Monterey *Herald*. The telephone rang and his city editor said that the decomposed body of a sea-serpent was washed up on the beach at Moss Landing, half-way around the Bay. Jimmy was to rush over and get pictures of it. He rushed, approached the evil-smelling monster from which the flesh was dropping. There was a note pinned to its head which said, "Don't worry about it, it's a basking shark. [Signed] Dr. Rolph Bolin of the Hopkins Marine Station." No doubt that Dr. Bolin acted kindly, for he loves true things; but his kindness was a blow to the people of Monterey. They so wanted it to be a sea-serpent. Even we hoped it would be. When sometimes a true sea-serpent, complete and undecayed, is found or caught, a shout of triumph will go through the world. "There, you see," men will say, "I knew they were there all the time. I just had a feeling they were there." Men really need sea-monsters in their personal oceans. And the Old Man of the Sea is one of these. In Monterey you can find many people who have seen him. Tiny Colletto has seen him close up and can draw a crabbed sketch of him. He is very large. He stands up in the water, three or four feet emerged above the waves, and watches an approaching boat until it comes too close, and then he sinks slowly out of sight. He looks somewhat like a tremendous diver, with large eyes and fur shaggily hanging from him. So far, he has not been photographed. When he is, probably Dr. Bolin will identify him and another beautiful story will be shattered. For this reason we rather hope he is never photographed, for if the Old

Man of the Sea should turn out to be some great malformed sea-lion, a lot of people would feel a sharp personal loss—a Santa Claus loss. And the ocean would be none the better for it. For the ocean, deep and black in the depths, is like the low dark levels of our minds in which the dream symbols incubate and sometimes rise up to sight like the Old Man of the Sea. And even if the symbol vision be horrible, it is there and it is ours. An ocean without its unnamed monsters would be like a completely dreamless sleep. Sparky and Tiny do not question the Old Man of the Sea, for they have looked at him. Nor do we question him because we know he is there. We would accept the testimony of these boys sufficiently to send a man to his death for murder, and we know they saw this monster and that they described him as they saw him.

We have thought often of this mass of sea-memory, or sea-thought, which lives deep in the mind. If one ask for a description of the unconscious, even the answer-symbol will usually be in terms of a dark water into which the light descends only a short distance. And we have thought how the human fetus has, at one stage of its development, vestigial gill-slits. If the gills are a component of the developing human, it is not unreasonable to suppose a parallel or concurrent mind or psyche development. If there be a life-memory strong enough to leave its symbol in vestigial gills, the preponderantly aquatic symbols in the individual unconscious might well be indications of a group psyche-memory which is the foundation of the whole unconscious. And what things must be there, what monsters, what enemies, what fear of dark and pressure, and of prey! There are numbers of examples wherein even invertebrates seem to remember and to react to stimuli no longer violent enough to cause the reaction. Perhaps, next to that of the sea, the strongest memory in us is that of the moon. But moon and sea and tide are one. Even now, the tide establishes a measurable, although minute, weight differential. For example, the steamship *Majestic* loses about fifteen pounds of its weight under a full moon.[1] According to a theory of George Darwin (son of Charles Darwin), in pre-Cambrian times, more than a thousand million years ago, the tides were tremendous; and the weight differential would have been correspondingly large. The moon-pull must have been the most

[1] Marmer, *The Tide*, 1926, p. 26.

important single environmental factor of littoral animals. Displacement and body weight then must certainly have decreased and increased tremendously with the rotation and phases of the moon, particularly if the orbit was at that time elliptic. The sun's reinforcement was probably slighter, relatively.

Consider, then, the effect of a decrease in pressure on gonads turgid with eggs or sperm, already almost bursting and awaiting the slight extra pull to discharge. (Note also the dehiscence of ova through the body walls of the polychaete worms. These ancient worms have their ancestry rooted in the Cambrian and they are little changed.) Now if we admit for the moment the potency of this tidal effect, we have only to add the concept of inherited psychic pattern we call "instinct" to get an inkling of the force of the lunar rhythm so deeply rooted in marine animals and even in higher animals and in man.

When the fishermen find the Old Man rising in the pathways of their boats, they may be experiencing a reality of past and present. This may not be a hallucination; in fact, it is little likely that it is. The interrelations are too delicate and too complicated. Tidal effects are mysterious and dark in the soul, and it may well be noted that even today the effect of the tides is more valid and strong and widespread than is generally supposed. For instance, it has been reported that radio reception is related to the rise and fall of Labrador tides,[2] and that there may be a relation between tidal rhythms and the recently observed fluctuations in the speed of light.[3] One could safely predict that all physiological processes correspondingly might be shown to be influenced by the tides, could we but read the indices with sufficient delicacy.

It appears that the physical evidence for this theory of George Darwin is more or less hypothetical, not in fact, but by interpretation, and that critical reasoning could conceivably throw out the whole process and with it the biologic connotations, because of unknown links and factors. Perhaps it should read the other way around. The animals themselves would seem to offer a striking confirmation to the tidal theory of cosmogony. One is almost forced to postulate some such theory if he would account causally for this

[2] *Science Supplement*, Vol. 80, No. 2069, p. 7, Aug. 24, 1934.
[3] *Science*, Vol. 81, No. 2091, p. 101, Jan. 25, 1935.

primitive impress. It would seem far-fetched to attribute the strong lunar effects actually observable in breeding animals to the present fairly weak tidal forces only, or to coincidence. There is tied up to the most primitive and powerful racial or collective instinct a rhythm sense or "memory" which affects everything and which in the past was probably more potent than it is now. It would at least be more plausible to attribute these profound effects to devastating and instinct-searing tidal influences active during the formative times of the early race history of organisms; and whether or not any mechanism has been discovered or is discoverable to carry on this imprint through the germ plasms, the fact remains that the imprint is there. The imprint is in us and in Sparky and in the ship's master, in the palolo worm, in mussel worms, in chitons, and in the menstrual cycle of women. The imprint lies heavily on our dreams and on the delicate threads of our nerves, and if this seems to come a long way from sea-serpents and the Old Man of the Sea, actually it has not come far at all. The harvest of symbols in our minds seems to have been planted in the soft rich soil of our pre-humanity. Symbol, the serpent, the sea, and the moon might well be only the signal light that the psycho-physiologic warp exists.

5

The evening came down on us and as it did the wind dropped but the tall waves remained, not topped with whitecaps any more. A few porpoises swam near and looked at us and swam away. The watches changed and we ate our first meal aboard, the cold wreckage of farewell snacks, and when our watch was done we were reluctant to go down to the bunks. We put on heavier coats and hung about the long bench where the helmsman sat. The little light on the compass card and the port and starboard lights were our outmost boundaries. Then we passed Point Sur and the waves flattened out into a ground-swell and increased in speed. Tony the master said, "Of course, it's always that way. The point draws the waves." Another might say, "The waves come greatly to the point," and in both statements there would be a good primitive exposition of the relation between giver and receiver. This relation would be through waves; wave to wave to wave, each of which is connected by torsion to its inshore fellow and touches it enough, although it has gone before, to be affected by its torsion. And so on and on to the shore, and to the point where the last wave, if you think from the sea, and the first if you think from the shore, touches and breaks. And it is important where you are thinking from.

The sharp, painful stars were out and bright enough to make the few whitecaps gleam against the dark surrounding water. From the wheel the little flag-jack on the peak stood against the course and swung back and forth over the horizon stars, blotting out each one as it passed. We tried to cover a star with the flag-jack and keep it covered, but this was impossible; no one could do that, not even Tony. But Tony, who knew his boat so well, could feel the yaw before it happened, could correct an error before it occurred. This is no longer reason or thought. One achieves the same feeling on a horse he knows well; one almost feels the horse's impulse in one's knees, and knows, but does not know, not only when the horse will shy, but the direction of his jump. The landsman, or the man who has been long ashore, is clumsy with the wheel, and his steering in a heavy sea is difficult. One grows tense on the wheel, particularly if someone like Tony is watching sardonically. Then keeping the compass card steady becomes impossible and the swing, a variable arc from two to ten degrees. And as weariness creeps up it is not uncommon to forget which way to turn the wheel to make the compass card swing back where you want it. The wheel turns only two ways, left or right. The fact of the lag, and the boat swinging rapidly so that a slow correcting allows it to pass the course and err on the other side, becomes a maddening thing when Tony the magnificent sits beside you. He does not correct you, he doesn't even speak. But Tony loves the truth, and the course is the truth. If the helmsman is off course he is telling a lie to Tony. And as the course projects, hypothetically, straight off the bow and around the world, so the wake drags out behind, a tattler on the conduct of the steersman. If one should steer mathematically perfectly, which is of course impossible, the wake will be a straight line; but even if, when drawn, it may have been straight, it bends to currents and to waves, and your true effort is wiped out. There is probably a unified-field hypothesis available in navigation as in all things. The internal factors would be the boat, the controls, the engine, and the crew, but chiefly the will and intent of the master, sub-headed with his conditioning experience, his sadness and ambitions and pleasures. The external factors would be the ocean with its bordering land, the waves and currents and the winds with their constant and varying effect in modifying the influence of the rudder against the changing tensions exerted on it.

If you steer *toward an object*, you cannot perfectly and indefinitely steer directly at it. You must steer to one side, or run it down; but you can steer exactly at a compass point, indefinitely. That does not change. Objects achieved are merely its fulfillment. In going toward a headland, for example, you can steer directly for it while you are at a distance, only changing course as you approach. Or you may set your compass course for the point and correct it by vision when you approach. The working out of the ideal into the real is here—the relationship between inward and outward, microcosm to macrocosm. The compass simply represents the ideal, present but unachievable, and sight-steering a compromise with perfection which allows your boat to exist at all.

In the development of navigation as thought and emotion—and

it must have been a slow, stumbling process frightening to its innovators and horrible to the fearful—how often must the questing mind have wished for a constant and unvarying point on the horizon to steer by. How simple if a star floated unchangeably to measure by. On clear nights such a star is there, but it is not trustworthy and the course of it is an arc. And the happy discovery of Stella Polaris—which, although it too shifts very minutely in an arc, is constant relatively—was encouraging. Stella Polaris will get you there. And so to the crawling minds Stella Polaris must have been like a very goddess of constancy, a star to love and trust.

What we have wanted always is an *unchangeable*, and we have found that only a compass point, a thought, an individual ideal, does not change—Schiller's and Goethe's *Ideal* to be worked out in terms of reality. And from such a thing as this, Beethoven writes a Ninth Symphony to Schiller's *Ode to Joy*.

A tide pool has been called a world under a rock, and so it might be said of navigation, "It is the world within the horizon."

Of steering, the external influences to be overcome are in the nature of oscillations; they are of short or long periods or both. The mean levels of the extreme ups and downs of the oscillations symbolize opposites in a Hegelian sense. No wonder, then, that in physics the symbol of oscillation, $\sqrt{-1}$, is fundamental and primitive and ubiquitous, turning up in every equation.

6

MARCH 12

In the morning we had come to the Santa Barbara Channel and the water was slick and gray, flowing in long smooth swells, and over it, close down, there hung a little mist so that the sea-birds flew in and out of sight. Then, breaking the water as though they swam in an obscure mirror, the porpoises surrounded us. They really came to us. We have seen them change course to join us, these curious animals. The Japanese will eat them, but rarely will Occidentals touch them. Of our crew, Tiny and Sparky, who loved to catch every manner of fish, to harpoon any swimming thing, would have nothing to do with porpoises. "They cry so," Sparky said, "when they are hurt, they cry to break your heart." This is rather a difficult thing to understand; a dying cow cries too, and a stuck pig raises his protesting voice piercingly and few hearts are broken by those cries. But a porpoise cries like a child in sorrow and pain. And we wonder whether the general seaman's real affection for porpoises might not be more complicated than the simple fear of hearing them cry. The nature of the animal might parallel certain traits in ourselves—the outrageous boastfulness of porpoises, their love of play, their joy in speed. We have watched them for many hours, making designs in the water, diving and rising and then seeming to turn over to see if they are watched. In bursts of speed they hump their backs and the beating tails take power from the whole body. Then they slow down and only the muscles near the tails are strained. They break the surface, and the blow-holes, like eyes, open and gasp in air and then close like eyes before they submerge. Suddenly they seem to grow tired of playing; the bodies hump up, the incredible tails beat, and instantly they are gone.

The mist lifted from the water but the oily slickness remained, and it was like new snow for keeping the impressions of what had happened there. Near to us was the greasy mess where a school of sardines had been milling, and on it the feathers of gulls which had come to join the sardines and, having fed hugely, had sat on the water and combed themselves in comfort. A Japanese liner passed us, slipping quickly through the smooth water, and for a long time we rocked in her wake. It was a long lazy day, and when the night came we passed the lights of Los Angeles with its many little dangling towns. The searchlights of the fleet at San Pedro combed the sea constantly, and one powerful glaring beam crept several miles and lay on us so brightly that it threw our shadows on the exhaust stack.

In the early morning before daylight we came into the harbor at San Diego, in through the narrow passage, and we followed the lights on a changing course to the pier. All about us war bustled, although we had no war; steel and thunder, powder and men—the men preparing thoughtlessly, like dead men, to destroy things. The planes roared over in formation and the submarines were quiet and ominous. There is no playfulness in a submarine. The military mind must limit its thinking to be able to perform its function at all. Thus, in talking with a naval officer who had won a target competition with big naval guns, we asked, "Have you thought what happens in a little street when one of your shells explodes, of the families torn to pieces, a thousand generations influenced when you signaled *Fire?*" "Of course not," he said. "Those shells travel so far that you couldn't possibly see where they land." And he was quite correct. If he could really see where they land and what they do, if he could really feel the power in his dropped hand and the waves radiating out from his gun, he would not be able to perform his function. He himself would be the weak point of his gun. But by not seeing, by insisting that it be a problem of ballistics and trajectory, he is a good gunnery officer. And he is too humble to take the responsibility for thinking. The whole structure of his world would be endangered if he permitted himself to think. The pieces must stick within their pattern or the whole thing collapses and the design is gone. We wonder whether in the present pattern the pieces are not straining to fall out of line; whether the paradoxes of our times are not finally mounting to a conclusion of ridiculousness that will make the whole structure collapse. For the paradoxes are becoming so great that leaders of people must be less and less intelligent to stand their own leadership.

The port of San Diego in that year was loaded with explosives and the means of transporting and depositing them on some enemy as yet undetermined. The men who directed this mechanism were true realists. They knew an enemy would emerge, and when one did, they had explosives to deposit on him.

In San Diego we filled the fuel tanks and the water tanks. We filled the icebox and took on the last perishable foods, bread and eggs and fresh meat. These would not last long, for when the ice was gone only the canned goods and the foods we could take from the sea would be available. We tied up to the pier all day and a night; got our last haircuts and ate broiled steaks.

This little expedition had become tremendously important to us; we felt a little as though we were dying. Strangers came to the pier and stared at us and small boys dropped on our deck like monkeys. Those quiet men who always stand on piers asked where we were going and when we said, "To the Gulf of California," their eyes melted with longing, they wanted to go so badly. They were like the men and women who stand about airports and railroad stations; they want to go away, and most of all they want to go away from themselves. For they do not know that they would carry their globes of boredom with them wherever they went. One man on the pier who wanted to participate made sure he would be allowed to cast us off, and he waited at the bow line for a long time. Finally he got the call and he cast off the bow line and ran back and cast off the stern line; then he stood and watched us pull away and he wanted very badly to go.

Below the Mexican border the water changes color; it takes on a deep ultramarine blue—a washtub bluing blue, intense and seeming to penetrate deep into the water; the fishermen call it "tuna water." By Friday we were off Point Baja. This is the region of the sea-turtle and the flying fish. Tiny and Sparky put out the fishing lines, and they stayed out during the whole trip.

Sparky Enea and Tiny Colletto grew up together in Monterey and they were bad little boys and very happy about it. It is said lightly that the police department had a special detail to supervise the growth and development of Tiny and Sparky. They are short and strong and nearly inseparable. An impulse seems to strike both of them at once. Let Tiny make a date with a girl and Sparky make a date with another girl—it then becomes necessary for Tiny, by connivance and trickery, to get Sparky's girl. But it is all right, since Sparky has been moving mountains to get Tiny's girl.

These two shared a watch, and on their watches we often went strangely off course and no one ever knew why. The compass had a way of getting out of hand so that the course invariably arced inshore. These two rigged the fishing lines with feathered artificial squid. Where the tackle was tied to the stays on either side, they looped the line and inset automobile inner tubes. For the tuna strikes so hard that something must give, and if the line does not break, the jaws tear off, so great is the combination of boat speed and tuna speed. The inner tube solves this problem by taking up the strain of the first great strike until direction and speed are equalized.

When Sparky and Tiny had the watch they took care of the fishing, and when the rubber tubes snapped and shook, one of them climbed down to take in the fish. If it were a large one, or a sharp-fighting fish, hysterical shrieks came from the fisherman. Whereupon the one left at the wheel came down to help and the wheel swung free. We wondered if this habit might not have caused the wonderful course we sailed sometimes. It is not beyond reason that coming back to the wheel, arguing and talking, they might have forgotten the set course and made one up almost as good. "Surely," they might think, "that is kinder and better than waking up the master to ask the course again, and five or ten degrees isn't so important when you aren't going far." If Tony loved the truth for itself, he was more than counterbalanced by Sparky and Tiny. They have little faith in truth, or, for that matter, in untruth. The police who had overseen their growing up had given them a nice appreciation of variables; they tested everything to find out whether it were true or not. In a like manner they tested the compass for a weakness they suspected was in it. And if Tony should say, "You are way off course," they could answer, "Well, we didn't hit anything, did we?"

7

MARCH 16

By two P.M. we were in the region of Magdalena Bay. The sea was still oily and smooth, and a light lacy fog lay on the water. The flying fish leaped from the forcing bow and flew off to right and left. It seemed, although this has not been verified, that they could fly farther at night than during the day. If, as is supposed, the flight is terminated when the flying fins dry in the air, this observation would seem to be justified, for at night they would not dry so quickly. Again, the whole thing might be a trick of our eyes. Often we played the searchlight on a fish in flight. The strangeness of light may have made the flight seem longer.

Tiny is a natural harpooner; often he had stood poised on the bow, holding the lance, but thus far nothing had appeared except porpoises, and these he would not strike. But now the sea-turtles began to appear in numbers. He stood for a long time waiting, and finally he drove his lance into one of them. Sparky promptly left the wheel, and the two of them pulled in a small turtle, about two and a half feet long. It was a tortoiseshell turtle.[1] Now we were able to observe the tender hearts of our crew. The small arrow-harpoon had penetrated the fairly soft shell, then turned sideways in the body. They hung the turtle to a stay where it waved its flippers helplessly and stretched its old wrinkled neck and gnashed its parrot beak. The small dark eyes had a quizzical pained look and a quantity of blood emerged from the pierced shell. Suddenly remorse seized Tiny; he wanted to put the animal out of its pain. He lowered the turtle to the deck and brought out an ax. With his first stroke he missed the animal entirely and sank the blade into the deck, but on his second stroke he severed the head from the body. And now a strange and terrible bit of knowledge came to Tiny; turtles are very hard to kill. Cutting off the head seems to

[1] *Eretmochelys imbricata* (Linn.). Nelson, but usually known as *Chelone imbricata.*

have little immediate effect. This turtle was as lively as it had been, and a large quantity of very red blood poured from the trunk of the neck. The flippers waved frantically and there was none of the constricting motion of a decapitated animal. We were eager to examine this turtle and we put Tiny's emotion aside for the moment. There were two barnacle bases on the shell and many hydroids which we preserved immediately. In the hollow beside the small tail were two pelagic crabs[2] of the square-fronted group, a male and a female; and from the way in which they hid themselves in the fold of turtle skin they seemed to be at home there. We were eager to examine the turtle's intestinal tract, both to find the food it had been eating and to look for possible tapeworms. To this end we sawed the shell open at the sides and opened the body cavity. From gullet to anus the digestive tract was crammed with small bright-red rock-lobsters[3]; a few of those nearest the gullet were whole enough to preserve. The gullet itself was lined with hard, sharp-pointed spikes, not of bone, but of a specialized tissue hard enough to macerate the small crustacea the turtle fed on. A curious peristalsis of the gullet (still observable, since even during dissection the reflexes were quite active) brought these points near together in a grinding motion and at the same time passed the increasingly macerated material downward toward the stomach. A good adaptation to food supply by structure, or perhaps vice versa. The heart continued to beat regularly. We removed it and placed it in a jar of salt water, where it continued to pulse for several hours; and twenty-four hours later, when it had apparently stopped, a touch with a glass rod caused it to pulse several times before it relaxed again. Tiny did not like this process of dissection. He wants his animals to die and be dead when he chops them; and when we cut up the muscular tissue, intending to cook it, and even the little cubes of white meat responded to touch, Tiny swore that he would give up sea-turtles and he never again tried to harpoon one. In his mind they joined the porpoises as protected animals. Probably he identified himself with the writhing tissue of the turtle and was unable to see it objectively.

The cooking was a failure. We boiled the meat, and later threw

[2] *Planes minutus* (Linn.).

[3] *Pleuroncodes planipes* Stimpson.

out the evil-smelling mess. (Subsequently, we discovered that one has to know how to cook a turtle.) But the turtle shell we wished to preserve. We scraped it as well as we could and salted it. Later we hung it deep in the water, hoping the isopods would clean it for us, but they never did. Finally we impregnated it with formaldehyde, then let it dry in the sun, and after all that we threw it away. It was never pretty and we never loved it.

During the night we crossed a school of bonito,[4] fast, clean-cut, beautiful fish of the mackerel family. The boys on watch caught five of them on the lines and during the process we got quite badly off course. We tried to take moving pictures of the color and of the color-pattern change which takes place in these fish during their death struggles. In the flurry when they beat the deck with their tails, the colors pulse and fade and brighten and fade again, until, when they are dead, a new pattern is visible. We wished to take color photographs of many of the animals because of the impossibility of retaining color in preserved specimens, and also because many animals, in fact most animals, have one color when they are alive and another when they are dead. However, none of us was expert in photography and we had a very mediocre success. The bonitos were good to eat, and Sparky fried big thick fillets for us.

That night we netted two small specimens of the northern flying fish.[5] Sparky, when we were looking at Barnhart's *Marine Fishes of Southern California*, saw a drawing of a lantern-fish entitled "*Monoceratias acanthias* after Gilbert" and he asked, "What's he after Gilbert for?"

This smooth blue water runs out of time very quickly, and a kind of dream sets in. Then a floating box cast overboard from some steamship becomes a fascinating thing, and it is nearly impossible not to bring the wheel over and go to pick it up. A new kind of porpoise began to appear, gray, where the northern porpoise had been dark brown. They were slim and very fast, the noses long and paddle-shaped. They move about in large schools, jumping out of the water and seeming to have a very good time. The abundance of life here gives one an exuberance, a feeling of fullness and richness. The playing porpoises, the turtles, the great schools

[4] *Sarda chiliensis* (Girard).
[5] *Cypselurus californicus.*

of fish which ruffle the water surface like a quick breeze, make for excitement. Sometimes in the distance we have seen a school of jumping tuna, and as they threw themselves clear of the water, the sun glittered on them for a moment. The sea here swarms with life, and probably the ocean bed is equally rich. Microscopically, the water is crowded with plankton. This is the tuna water—life water. It is complete from plankton to gray porpoises. The turtle was complete with the little almost-commensal crab living under his tail and with barnacles and hydroids riding on his back. The pelagic rock-lobsters[6] littered the ocean with red spots. There was food everywhere. Everything ate everything else with a furious exuberance.

About five P.M. on the sixteenth, seventy miles north of Point Lazaro, we came upon hosts of the red rock-lobsters on the surface, brilliant red and beautiful against the ultramarine of the water. There was no protective coloration here—a greater contrast could not have been chosen. The water seemed almost solid with the little red crustacea, called "*langustina*" by the Mexicans. According to Stimpson, on March 8, 1859, a number of them were thrown ashore at Monterey in California, many hundreds of miles from their usual range. It was probably during one of those queer cycles when the currents do amazing things. We idled our engine and crept slowly along catching up the *langustina* in dip-nets. We put them in white porcelain pans and took some color moving pictures of them—some of the few good moving pictures, incidentally, made during the whole trip. In the pans we saw that these animals do not swim rapidly, but rather wriggle and crawl through the water. Finally, we immersed them in fresh water and when they were dead, preserved them in alcohol, which promptly removed their brilliant color.

[6] *Pleuroncodes.*

8

MARCH 17

At two A.M. we passed Point Lazaro, one of the reputedly dangerous places of the world, like Cedros Passage, or like Cape Horn, where the weather is always bad even when it is good elsewhere. There is a sense of relief when one is safely past these half-mythical places, for they are not only stormy but treacherous, and again the atavistic fear arises—the Scylla-Charybdis fear that made our ancestors people such places with monsters and enter them only after prayer and propitiation. It was only reasonably rough when we passed, and immediately south the water was very calm. About five in the morning we came upon an even denser concentration of the little red *Pleuroncodes,* and we stopped again and took a great many of them. While we netted the *langustina,* a skipjack struck the line and we brought him in and had him for breakfast. During the meal we said the fish was *Katsuwonus pelamis,* and Sparky said it was a skipjack because he was eating it and he was quite sure he would not eat *Katsuwonus pelamis* ever. A few hours later we caught two small dolphins,[1] startlingly beautiful fish of pure gold, pulsing and fading and changing colors. These fish are very widely distributed.

We were coming now toward the end of our day-and-night running; the engine had never paused since we left San Diego except for idling the little time while we took the *langustina.* The coastline of the Peninsula slid along, brown and desolate and dry with strange flat mountains and rocks torn by dryness, and the heat shimmer hung over the land even in March. Tony had kept us well offshore, and only now we approached closer to land, for we would arrive at Cape San Lucas in the night, and from then on we planned to run only in the daytime. Some collecting stations we had projected, like Pulmo Reef and La Paz and Angeles Bay, but except for those, we planned to stop wherever the shore looked interesting. Even this little trip of ninety hours, though, had grown long,

[1] *Coryphaena equisetis* Linn.

and we were glad to be getting to the end of it. The dry hills were red gold that afternoon and in the night no one left the top of the deckhouse. The Southern Cross was well above the horizon, and the air was warm and pleasant. Tony spent a long time in the galley going over the charts. He had been to Cape San Lucas once before. Around ten o'clock we saw the lighthouse on the false cape. The night was extremely dark when we rounded the end; the great tall rocks called "The Friars" were blackly visible. The *Coast Pilot* spoke of a light on the end of the San Lucas pier, but we could see no light. Tony edged the boat slowly into the dark harbor. Once a flashlight showed for a moment on the shore and then went out. It was after midnight, and of course there would be no light in a Mexican house at such a time. The searchlight on our deckhouse seemed to be sucked up by the darkness. Sparky on the bow with the leadline found deep water, and we moved slowly in, stopping and drifting and sounding. And then suddenly there was the beach, thirty feet away, with little waves breaking on it, and still we had eight fathoms on the lead. We backed away a little and dropped the anchor and waited until it took a firm grip. Then the engine stopped, and we sat for a long time on the deckhouse. The sweet smell of the land blew out to us on a warm wind, a smell of sand verbena and grass and mangrove. It is so quickly forgotten, this land smell. We know it so well on shore that the nose forgets it, but after a few days at sea the odor memory pattern is lost so that the first land smell strikes a powerful emotional nostalgia, very sharp and strangely dear.

In the morning the black mystery of the night was gone and the little harbor was shining and warm. The tuna cannery against the gathering rocks of the point and a few houses along the edge of the beach were the only habitations visible. And with the day came the answer to the lightlessness of the night before. The *Coast Pilot* had not been wrong. There is indeed a light on the end of the cannery pier, but since the electricity is generated by the cannery engine, and since the cannery engine runs only in the daytime, so the light burns only in the daytime. With the arrived day, this light came on and burned bravely until dusk, when it went off again. But the *Coast Pilot* was absolved, it had not lied. Even Tony, who had been a little bitter the night before, was forced to revise his first fierceness. And perhaps it was a lesson to Tony in exact thinking, like those carefully worded puzzles in joke books; the *Pilot* said a light burned—it only neglected to say when, and we ourselves supplied the fallacy.

The great rocks on the end of the Peninsula are almost literary. They are a fitting Land's End, standing against the sea, the end of a thousand miles of peninsula and mountain. Good Hope is this way too, and perhaps we take some of our deep feelings of termination from these things, and they make our symbols. The Friars stood high and protective against an interminable sea.

Clavigero, a Jesuit monk, came to the Point and the Peninsula over two hundred years ago. We quote from the Lake and Gray translation of his history of Lower California,[2] page fifteen: "This Cape is its southern terminus, the Red River [Colorado] is the eastern limit, and the harbor of San Diego, situated at 33 degrees north latitude and about 156 degrees longitude, can be called its western limit. To the north and the northeast it borders on the countries of barbarous nations little known on the coasts and not at all in the interior. To the west it has the Pacific Sea and on the east the Gulf of California, already called the Red Sea because of its similarity to the Red Sea, and the Sea of Cortés, named in honor of the famous conqueror of Mexico who had it discovered and who navigated it. The length of the Peninsula is about 10 degrees, but its width varies from 30 to 70 miles and more.

"The name, California," Clavigero goes on, "was applied to a single port in the beginning, but later it was extended to mean all the Peninsula. Some geographers have even taken the liberty of comprising under this denomination New Mexico, the country of the Apaches, and other regions very remote from the true California and which have nothing to do with it."

Clavigero says of its naming, "The origin of this name is not known, but it is believed that the conqueror, Cortés, who pretended to have some knowledge of Latin, named the harbor, where he put in, '*Callida fornax*' because of the great heat which he felt there; and that either he himself or some one of the many persons who accompanied him formed the name California from these two words. If this conjecture be not true, it is at least credible."

We like Clavigero for these last words. He was a careful man.

[2] Stanford University Press, 1937.

The observations set down in his history of Baja California are surprisingly correct, and if not all true, they are at least all credible. He always gives one his choice. Perhaps his Jesuit training is never more evident than in this. "If you believe this," he says in effect, "perhaps you are not right, but at least you are not a fool."

Lake and Gray include an interesting footnote in their translation. "The famous corsair, Drake, called California 'New Albion' in honor of his native land. Father Scherer, a German Jesuit, and M. de Fer, a French geographer, used the name 'Carolina Island' to designate California, which name began to be used in the time of Charles II, King of Spain, when that Peninsula was considered an island, but these and other names were soon forgotten and that given it by the conqueror, Cortés, prevailed."

And in a second footnote, Lake and Gray continue, "We shall add the opinion of the learned ex-Jesuit, Don José Campoi, on the etymology of the name, 'California,' or 'Californias' as others say. This Father believes that the said name is composed of the Spanish word '*Cala*' which means a small cove of the sea, and the latin word '*fornix*' which means an arch; because there is a small cove at the cape of San Lucas on the western side of which there overhangs a rock pierced in such a way that in the upper part of that great opening is seen an arch formed so perfectly that it appears made by human skill. Therefore Cortés, noticing the cove and arch, and understanding Latin, probably gave to that port the name 'California' or *Cala-y-fornix*, speaking half Spanish and half Latin.

"To these conjectures we could add a third one, composed of both, by saying that the name is derived from *Cala,* as Campoi thinks, and *fornax,* as the author believes, because of the cove, and the heat which Cortés felt there, and that the latter might have called that place *Cala, y fornax.*" This ends the footnote.

Our feeling about this, and all the erudite discussion of the origin of this and other names, is that none of these is true. Names attach themselves to places and stick or fall away. When men finally go to live in Antarctica it is unlikely that they will ever speak of the Rockefeller Mountains or use the names designated by breakfast food companies. More likely a name emerges almost automatically from a place as well as from a man and the relationship between name and thing is very close. In the naming of places in the West this has seemed apparent. In this connection there are two examples: in the Sierras there are two little mountains which were called by the early settlers "Maggie's Bubs." This name was satisfactory and descriptive, but it seemed vulgar to later and more delicate lovers of nature, who tried to change the name a number of times and failing, in usage at least, finally surrendered and called them "The Maggies," explaining that it was an Indian name. In the same way Dog ——— Point (and I am delicate only for those same nature lovers) has had finally to be called in print "The Dog." It does not look like a dog, but it does look like that part of a dog which first suggested its name. However, anyone seeing this point immediately reverts to the designation which was anatomically accurate and strangely satisfying to the name-giving faculty. And this name-giving faculty is very highly developed and deeply rooted in our atavistic magics. To name a thing has always been to make it familiar and therefore a little less dangerous to us. "Tree" the abstract may harbor some evil until it has a name, but once having a name one can cope with it. A tree is not dangerous, but the forest is. Among primitives sometimes evil is escaped by never mentioning the name, as in Malaysia, where one never mentions a tiger by name for fear of calling him. Among others, as even among ourselves, the giving of a name establishes a familiarity which renders the thing impotent. It is interesting to see how some scientists and philosophers, who are an emotional and fearful group, are able to protect themselves against fear. In a modern scene, when the horizons stretch out and your philosopher is likely to fall off the world like a Dark Ages mariner, he can save himself by establishing a taboo-box which he may call "mysticism" or "supernaturalism" or "radicalism." Into this box he can throw all those thoughts which frighten him and thus be safe from them. But in geographic naming it seems almost as though the place contributed something to its own name. As Tony says, "The point draws the waves"—we say, "The place draws the name." It doesn't matter what California means; what does matter is that with all the names bestowed upon this place, "California" has seemed right to those who have seen it. And the meaningless word "California" has completely routed all the "New Albions" and "Carolinas" from the scene.

The strangest case of nicknaming we know concerns a man whose first name is Copeland. In three different parts of the country where he has gone, not knowing anyone, he has been called first "Copenhagen" and then "Hagen." This has happened automatically. He is Hagen. We don't know what quality of Hagenness he has, but there must be some. Why not "Copen" or "Cope"? It is never that. He is invariably Hagen. This, we realize, has become mystical, and anyone who wishes may now toss the whole thing into his taboo-box and slam the lid down on it.

The tip of the Cape at San Lucas, with the huge gray Friars standing up on the end, has behind the rocks a little beach which is a small boy's dream of pirates. It seems the perfect place to hide and from which to dart out in a pinnace on the shipping of the world; a place to which to bring the gold bars and jewels and beautiful ladies, all of which are invariably carried by the shipping of the world. And this little beach must so have appealed to earlier men, for the names of pirates are still in the rock, and the pirate ships did dart out of here and did come back. But now in back of the Friars on the beach there is a great pile of decaying hammerhead sharks, the livers torn out and the fish left to rot. Some day, and that soon, the more mature piracy which has abandoned the pinnace for the coast gun will stud this point with gray monsters and will send against the shipping of the Gulf, not little bands of ragged men, but projectiles filled with TNT. And from that piracy no jewels or beautiful ladies will come back to the beach behind the rocks.

On that first morning we cleaned ourselves well and shaved while we waited for the Mexican officials to come out and give us the right to land. They were late in coming, for they had to find their official uniforms, and they too had to shave. Few boats put in here. It would not be well to waste the occasion of the visit of even a fishing boat like ours. It was noon before the well-dressed men in their sun helmets came down to the beach and were rowed out to us. They were armed with the .45-caliber automatics which everywhere in Mexico designate officials. And they were armed also with the courtesy which is unique in official Mexico. No matter what they do to you, they are nice about it. We soon learned the routine in other ports as well as here. Everyone who has or can borrow a uniform comes aboard—the collector of customs in a washed and shiny uniform; the business agent in a business suit having about him what Tiny calls "a double-breasted look"; then soldiers if there are any; and finally the Indians, who row the boat and rarely have uniforms. They come over the side like ambassadors. We shake hands all around. The galley has been prepared: coffee is ready and perhaps a drop of rum. Cigarettes are presented and then comes the ceremonial of the match. In Mexico cigarettes are cheap, but matches are not. If a man wishes to honor you, he lights your cigarette, and if you have given him a cigarette, he must so honor you. But having lighted your cigarette and his, the match is still burning and not being used. Anyone may now make use of this match. On a street, strangers who have been wishing for a light come up quickly and light from your match, bow, and pass on.

We were impatient for the officials, and this time we did not have to wait long. It developed that the Governor of the southern district had very recently been to Cape San Lucas and just before that a yacht had put in. This simplified matters, for, having recently used them, the officials knew exactly where to find their uniforms, and, having found them, they did not, as sometimes happens, have to send them to be laundered before they could come aboard. About noon they trooped to the beach, scattering the pigs and Mexican vultures which browsed happily there. They filled the rowboat until the gunwales just missed dipping, and majestically they came alongside. We conducted the ceremony of clearing with some dignity, for if we spoke to them in very bad Spanish, they in turn honored us with very bad English. They cleared us, drank coffee, smoked, and finally left, promising to come back. Much as we had enjoyed them, we were impatient, for the tide was dropping and the exposed rocks looked very rich with animal life.

All the time we were indulging in courtliness there had been light gunfire on the cliffs, where several men were shooting at black cormorants; and it developed that everyone in Cape San Lucas hates cormorants. They are the flies in a perfect ecological ointment. The cannery cans tuna; the entrails and cuttings of the tuna are thrown into the water from the end of the pier. This refuse brings in schools of small fish which are netted and used for bait to catch tuna. This closed and tight circle is interfered with by the cormorants, who try to get at the bait-fish. They dive and catch fish, but also they drive the schools away from the pier out of easy reach of the bait-men. Thus they are considered interlopers, radicals, subversive forces against the perfect and God-set balance on Cape San Lucas. And they are rightly slaughtered, as all radicals should be. As one of our number remarked, "Why, pretty soon they'll want to vote."

Finally we could go. We unpacked the Hansen Sea-Cow and fastened it on the back of the skiff. This was our first use of the Sea-Cow. The shore was very close and we were able just by pulling on the starter rope to spin the propeller enough to get us to shore. The Sea-Cow did not run that day but it seemed to enjoy having its flywheel spun.

The shore-collecting equipment usually consisted of a number of small wrecking bars; wooden fish-kits with handles; quart jars with screw caps; and many glass tubes. These tubes are invaluable for small and delicate animals: the chance of bringing them back uninjured is greatly increased if each individual, or at least only a few of like species, are kept in separate containers. We filled our pockets with these tubes. The soft animals must never be put in the same container with any of the livelier crabs, for these, when restrained or inhibited in any way, go into paroxysms of rage and pinch everything at random, even each other; sometimes even themselves.

The exposed rocks had looked rich with life under the lowering tide, but they were more than that: they were ferocious with life. There was an exuberant fierceness in the littoral here, a vital competition for existence. Everything seemed speeded-up; starfish and urchins were more strongly attached than in other places, and many of the univalves were so tightly fixed that the shells broke before the animals would let go their hold. Perhaps the force of the great surf which beats on this shore has much to do with the tenacity of the animals here. It is noteworthy that the animals, rather than deserting such beaten shores for the safe cove and protected pools, simply increase their toughness and fight back at the sea with a kind of joyful survival. This ferocious survival quotient excites us and makes us feel good, and from the crawling, fighting, resisting qualities of the animals, it almost seems that they are excited too.

We collected down the littoral as the water went down. We didn't seem to have time enough. We took samples of everything that came to hand. The uppermost rocks swarmed with Sally Lightfoots, those beautiful and fast and sensitive crabs. With them were white periwinkle snails. Below that, barnacles and Purpura snails; more crabs and many limpets. Below that many serpulids—attached worms in calcareous tubes with beautiful purple floriate heads. Below that, the multi-rayed starfish, *Heliaster kubiniji* of Xanthus. With *Heliaster* were a few urchins, but not many, and they were so placed in crevices as to be hard to dislodge. Several resisted the steel bar to the extent of breaking—the mouth remaining tight to the rock while the shell fell away. Lower still there were to be seen swaying in the water under the reefs the dark gorgonians, or sea-fans. In the lowest surf-levels there was a brilliant gathering of the moss animals known as bryozoa; flatworms; flat crabs; the large sea-cucumber[3]; some anemones; many sponges of two types, a smooth, encrusting purple one, the other erect, white, and calcareous. There were great colonies of tunicates, clusters of tiny individuals joined by a common tunic and looking so like the sponges that even a trained worker must await the specialist's determination to know whether his find is sponge or tunicate. This is annoying, for the sponge being one step above the protozoa, at the bottom of the evolutionary ladder, and the tunicate near the top, bordering the vertebrates, your trained worker is likely to feel that a dirty trick has been played upon him by an entirely too democratic Providence.

We took many snails, including cones and murexes; a small red tectibranch (of a group to which the sea-hares belong); hydroids; many annelid worms; and a red pentagonal starfish.[4] There were the usual hordes of hermit crabs, but oddly enough we saw no chitons (sea-cradles), although the region seemed ideally suited to them.

We collected in haste. As the tide went down we kept a little ahead of it, wading in rubber boots, and as it came up again it drove us back. The time seemed very short. The incredible beauty of the tide pools, the brilliant colors, the swarming species ate up the time. And when at last the afternoon surf began to beat on the littoral and covered it over again, we seemed barely to have started. But the buckets and jars and tubes were full, and when we stopped we discovered that we were very tired.

Our collecting ends were different from those ordinarily entertained. In most cases at the present time, collecting is done by men who specialize in one or more groups. Thus, one man interested in

[3] *Holothuria lubrica.*

[4] *Oreaster.*

hydroids will move out on a reef, and if his interest is sharp enough, he will not even see other life forms about him. For him, the sponge is something in the way of his hydroids. Collecting large numbers of animals presents an entirely different aspect and makes one see an entirely different picture. Being more interested in distribution than in individuals, we saw dominant species and changing sizes, groups which thrive and those which recede under varying conditions. In a way, ours is the older method, somewhat like that of Darwin on the *Beagle*. He was called a "naturalist." He wanted to see everything, rocks and flora and fauna; marine and terrestrial. We came to envy this Darwin on his sailing ship. He had so much room and so much time. He could capture his animals and keep them alive and watch them. He had years instead of weeks, and he saw so many things. Often we envied the inadequate transportation of his time—the *Beagle* couldn't get about rapidly. She moved slowly along under sail. And we can imagine that young Darwin, probably in a bos'n's chair hung over the side, with a dip-net in his hands, scooping up jellyfish. When he went inland, he rode a horse or walked. This is the proper pace for a naturalist. Faced with all things he cannot hurry. We must have time to think and to look and to consider. And the modern process—that of looking quickly at the whole field and then diving down to a particular—was reversed by Darwin. Out of long long consideration of the parts he emerged with a sense of the whole. Where we wished for a month at a collecting station and took two days, Darwin stayed three months. Of course he could see and tabulate. It was the pace that made the difference. And in the writing of Darwin, as in his thinking, there is the slow heave of a sailing ship, and the patience of waiting for a tide. The results are bound up with the pace. We *could* not do this even if we could. We have thought in this connection that the speed and tempo and tone of modern writing might be built on the nervous clacking of a typewriter; that the brittle jerky thinking of the present might rest on the brittle jerky curricula of our schools with their urge to "turn them out." To turn them out. They use the phrase in speeches; turn them out to what? And the young biologists tearing off pieces of their subject, tatters of the life forms, like sharks tearing out hunks of a dead horse, looking at them, tossing them away. This is neither a good nor a bad method; it is simply the one of our time. We can look with longing back to Charles Darwin, staring into the water over the side of the sailing ship, but for us to attempt to imitate that procedure would be romantic and silly. To take a sailing boat, to fight tide and wind, to move four hundred miles on a horse when we could take a plane, would be not only ridiculous but ineffective. For we first, before our work, are products of our time. We might produce a philosophical costume piece, but it would be completely artificial. However, we can and do look on the measured, slow-paced accumulation of sight and thought of the Darwins with a nostalgic longing.

Even our boat hurried us, and while the Sea-Cow would not run, it had nevertheless infected us with the idea of its running. Six weeks we had, and no more. Was it a wonder that we collected furiously; spent every low-tide moment on the rocks, even at night? And in the times between low tides we kept the bottom nets down and the lines and dip-nets working. When the charter was up, we would be through. How different it had been when John Xantus was stationed in this very place, Cape San Lucas, in the sixties. Sent down by the United States Government as a tidal observer, but having lots of time, he collected animals for our National Museum. The first fine collections of Gulf forms came from Xantus. And we do not feel that we are injuring his reputation, but rather broadening it, by repeating a story about him. Speaking to the manager of the cannery at the Cape, we remarked on what a great man Xantus had been. Where another would have kept his tide charts and brooded and wished for the Willard Hotel, Xantus had collected animals widely and carefully. The manager said, "Oh, he was even better than that." Pointing to three little Indian children he said, "Those are Xantus's great-grandchildren," and he continued, "In the town there is a large family of Xantuses, and a few miles back in the hills you'll find a whole tribe of them." There were giants in the earth in those days.

We wonder what modern biologist, worried about titles and preferment and the gossip of the Faculty Club, would have the warmth and breadth, or even the fecundity for that matter, to leave a "whole tribe of Xantuses." We honor this man for all his activities. He at least was one who literally did proliferate in all directions.

Many people have spoken at length of the Sally Lightfoots. In fact, everyone who has seen them has been delighted with them.

The very name they are called by reflects the delight of the name. These little crabs, with brilliant cloisonné carapaces, walk on their tiptoes. They have remarkable eyes and an extremely fast reaction time. In spite of the fact that they swarm on the rocks at the Cape, and to a less degree inside the Gulf, they are exceedingly hard to catch. They seem to be able to run in any one of four directions; but more than this, perhaps because of their rapid reaction time, they appear to read the mind of their hunter. They escape the long-handled net, anticipating from what direction it is coming. If you walk slowly, they move slowly ahead of you in droves. If you hurry, they hurry. When you plunge at them, they seem to disappear in little puffs of blue smoke—at any rate, they disappear. It is impossible to creep up on them. They are very beautiful, with clear brilliant colors, reds and blues and warm browns. We tried for a long time to catch them. Finally, seeing fifty or sixty in a big canyon of rock, we thought to outwit them. Surely we were more intelligent, if slower, than they. Accordingly, we pitted our obviously superior intelligence against the equally obvious physical superiority of Sally Lightfoot. Near the top of the crevice a boulder protruded. One of our party, taking a secret and circuitous route, hid himself behind this boulder, net in hand. He was completely concealed even from the stalk eyes of the crabs. Certainly they had not seen him go there. The herd of Sallys drowsed on the rocks in the lower end of the crevice. Two more of us strolled in from the seaward side, nonchalance in our postures and ingenuousness on our faces. One might have thought that we merely strolled along in a contemplation which severely excluded Sally Lightfoots. In time the herd moved ahead of us, matching our nonchalance. We did not hurry, they did not hurry. When they passed the boulder, helpless and unsuspecting, a large net was to fall over them and imprison them. But they did not know that. They moved along until they were four feet from the boulder, and then as one crab they turned to the right, climbed up over the edge of the crevice and down to the sea again.

Man reacts peculiarly but consistently in his relationship with Sally Lightfoot. His tendency eventually is to scream curses, to hurl himself at them, and to come up foaming with rage and bruised all over his chest. Thus, Tiny, leaping forward, slipped and fell and hurt his arm. He never forgot nor forgave his enemy. From then on he attacked Lightfoots by every foul means he could contrive (and a training in Monterey street fighting had equipped him well for this kind of battle). He hurled rocks at them; he smashed at them with boards; and he even considered poisoning them. Eventually we did catch a few Sallys, but we think they were the halt and the blind, the simpletons of their species. With reasonably well-balanced and non-neurotic Lightfoots we stood no chance.

We came back to the boat loaded with specimens, and immediately prepared to preserve them. The square, enameled pans were laid out on the hatch, the trays and bowls and watchglasses (so called because at one time actual watch-crystals were used). The pans and glasses were filled with fresh sea water, and into them we distributed the animals by families—all the crabs in one, anemones in another, snails in another, and delicate things like flatworms and hydroids in others. From this distribution it was easier to separate them finally by species.

9

When the catch was sorted and labeled, we went ashore to the cannery and later drove with Chris, the manager, and Señor Luis, the port captain, to the little town of San Lucas. It was a sad little town, for a winter storm and a great surf had wrecked it in a single night. Water had driven past the houses, and the streets of the village had been a raging river. "Then there were no roofs over the heads of the people," Señor Luis said excitedly. "Then the babies cried and there was no food. Then the people suffered."

The road to the little town, two wheel-ruts in the dust, tossed us about in the cannery truck. The cactus and thorny shrubs ripped at the car as we went by. At last we stopped in front of a mournful *cantina* where morose young men hung about waiting for something to happen. They had waited a long time—several generations—for something to happen, these good-looking young men. In their eyes there was a hopelessness. The storm of the winter had been discussed so often that it was sucked dry. And besides, they all knew the same things about it. Then we happened to them. The truck pulled up to the *cantina* door and we—strangers, foreigners—stepped out, as disorderly-looking a group as had ever come to their *cantina*. Tiny wore a Navy cap of white he had traded for, he said, in a washroom in San Diego. Tony still had his snap-brim felt. There were yachting caps and sweaters, and jeans stiff with fish blood. The young men stirred to life for a little while, but we were not enough. The flood had been much better. They relapsed again into their gloom.

There is nothing more doleful than a little *cantina*. In the first place it is inhabited by people who haven't any money to buy a drink. They stand about waiting for a miracle that never happens: the angel with golden wings who settles on the bar and orders drinks for everyone. This never happens, but how are the sad handsome young men to know it never will happen? And suppose it did happen and they were somewhere else? And so they lean against the wall; and when the sun is high they sit down against the wall. Now and then they go away into the brush for a while, and they go to their little homes for meals. But that is an impatient time, for the golden angel might arrive. Their faith is not strong, but it is permanent.

We could see that we did not greatly arouse them. The *cantina* owner promptly put his loudest records on the phonograph to force a gaiety into this sad place. But he had Carta Blanca beer and (at the risk of a charge that we have sold our souls to this brewery) we love Carta Blanca beer. There was no ice, no electric lights, and the gasoline lanterns hissed and drew the bugs from miles away. The cockroaches in their hordes rushed in to see what was up. Big, handsome cockroaches, with almost human faces. The loud music only made us sadder, and the young men watched us. When we lifted a split of beer to our lips the eyes of the young men rose with our hands, and even the cockroaches lifted their heads. We couldn't stand it. We ordered beer all around, but it was too late. The young men were too far gone in sorrow. They drank their warm beer sadly. Then we bought straw hats, for the sun is deadly here. There should be a kind of ridiculous joy in buying a floppy hat, but those young men, so near to tears, drained even that joy. Their golden angel had come, and they did not find him good. We felt rather as God would feel when, after all the preparation of Paradise, all the plannings for eternities of joy, all the making and tuning of harps, the street-paving with gold, and the writing of hosannas, at last He let in the bleacher customers and they looked at the heavenly city and wished to be again in Brooklyn. We told funny stories, knowing they wouldn't be enjoyed, tiring of them ourselves before the point was reached. Nothing was fun in that little *cantina*. We started back for the boat. I think those young men were glad to see us go; because once we were gone, they could begin to build us up, but present, we inhibited their imaginations.

At the bar Chris told us of a native liquor called *damiana,* made from an infusion of a native herb, and not much known outside of Baja California. Chris said it was an aphrodisiac, and told some interesting stories to prove it. We felt a scientific interest in his stories, and bought a bottle of *damiana,* intending to subject it to certain tests under laboratory conditions. But the customs officials of San Diego took it away from us, not because of its romantic aspect, but because it had alcohol in it. Thus we were never able to give it a truly scientific testing. We think we were going to use it on a white rat. Tiny said he didn't want any such stuff getting in his way when he felt lustful.

There doesn't seem to be a true aphrodisiac; there are excitants like cantharides, and physical aids to the difficulties of psychic traumas, like yohimbine sulphate; there are strong protein foods like *bêche-de-mer* and the gonads of sea-urchins, and the much over-rated oyster; even chiles, with their irritating qualities, have some effect, but there seems to be no true aphrodisiac, no sweet essence of that goddess to be taken in a capsule. A certain young person said once that she found sexual intercourse an aphrodisiac; certainly it is the only good one.

So many people are interested in this subject but most of them are forced to pretend they are not. A man, for his own ego's sake, must, publicly at least, be over-supplied with libido. But every doctor knows so well the "friend of the client" who needs help. He is the same "friend" who has gonorrhea, the same "friend" who needs the address of an abortionist. This elusive friend—what will we not do to help him out of his difficulties; the nights we spend sleepless, worrying about him! He is interested in an aphrodisiac; we must try to find him one. But the *damiana* we brought back for our "friend" possibly just now is in the hands of the customs officials in San Diego. Perhaps they too have a friend. Since we suggested the qualities of *damiana* to them, it is barely possible that this fascinating liquor has already been either devoted to a friend or even perhaps subjected to a stern course of investigation under laboratory conditions.

We have wondered about the bawdiness this book must have if it is to be true. Bawdiness, vulgarity—call it what you will—is such a relative matter, so much a matter of attitude. A man we know once long ago worked for a wealthy family in a country place. One morning one of the cows had a calf. The children of the house went down with him to watch her. It was a good normal birth, a perfect presentation, and the cow needed no help. The children asked questions and he answered them. And when the emerged head cleared through the sac, the little black muzzle appeared, and the first breath was drawn, the children were fascinated and awed. And this was the time for their mother to come screaming down on the vulgarity of letting the children see the birth. This "vulgarity" had given them a sense of wonder at the structure of life, while the mother's propriety and gentility supplanted that feeling with dirtiness. If the reader of this book is "genteel," then this is a very vulgar book, because the animals in a tide pool have two major preoccupations: first, survival, and second, reproduction. They reproduce all over the place. We could retire into obscure phrases or into Greek or Latin. This, for some reason, protects the delicate. In an earlier time biologists made their little jokes that way, as in the naming of the animals. But some later men found their methods vulgar. Verrill, in *The Actinaria of the Canadian Arctic Expeditions,* broke out in protest. He cries, "Prof. McMurrich has endeavored to restore for this species a name (*senilis*) used by Linnaeus for a small indeterminable species very imperfectly described in 1761. . . . The description does not in the least apply to this species. He described the thing as the size of the last joint of a finger, sordid, rough, with a sub-coriaceous tunic. Such a description could not possibly apply to this soft and smooth species . . . but it would be mere guesswork to say what species he had in view. . . . Moreover, aside from this uncertainty, most modern writers have rejected most of the Linnaean names of actinians on account of their obscenity or indecency. All this confusion shows the impossibility of fixing the name, even if it were not otherwise objectionable. It should be forgotten or ignored, like the generic names used by Linnaeus in 1761, and by some others of that period, for species of Actinia. Their indecent names were usually the Latinized forms of vulgar names used by fishermen, some of which are still in use among the fishermen of our own coasts, for similar things."

This strange attempt to "clean up" biology will have, we hope, no effect whatever. We at least have kept our vulgar sense of wonder. We are no better than the animals; in fact in a lot of ways we aren't as good. And so we'll let the book fall as it may.

We left the truck and walked through the sandy hills in the night, and in this latitude the sky seemed very black and the stars very white. Already the smell of the land was gone from our noses, for we were used to the smell of vegetation again. The beer was warm in us and pleasant, and the air had a liquid warmth that was really there without the beer, for we tested it later. In the brush beside the track there was a little heap of light, and as we came closer to it we saw a rough wooden cross lighted indirectly. The cross-arm was bound to the staff with a thong, and the whole cross seemed to glow, alone in the darkness. When we came close we saw that a kerosene can stood on the ground and that in it was a candle which threw its feeble light upward on the cross. And our companion told us how a man had come from a fishing boat, sick and weak and tired. He tried to get home, but at this spot he fell down and died. And his family put the little cross and the candle there to mark the place. And eventually they would put up a stronger cross. It seems good to mark and to remember for a little while the place where a man died. This is his one whole lonely act in all his life. In every other thing, even in his birth, he is bound close to others, but the moment of his dying is his own. And in nearly all of Mexico such places are marked. A grave is quite a different thing. Here one's family boasts, or lies, or excuses, in material of elegance and extravagance. But that is a family or a social matter, not the dead man's own at all. The unmarked cross and the secret light are his; almost a reflection of the last piercing loneliness that comes into a dying man's eyes.

From a few feet away the cross seemed to flicker unsubstantially with a small yellow light, seemed to be almost a memory while we saw it. And the man who tried to get home and crawled this far—we never knew his name but he stays in our memory too, for some reason—a supra-personal being, a slow, painful symbol and a pattern of his whole species which tries always from generation to generation, man and woman, which struggles always to get home but never quite makes it.

We came back to the pier and got into our little boat. The Sea-Cow of course would not start, it being night time, so we rowed out to the *Western Flyer*. Before we started, by some magic, there on the end of the pier stood the sad beautiful young men watching us. They had not moved; some jinni had picked them up and transported them and set them down. They watched us put out into the darkness toward our riding lights, and then we suppose they were whisked back again to the *cantina,* where the proprietor was putting the records away and feeling with delicate thumbs the dollar bills we had left. On the pier no light burned, for the engine had stopped at sundown. We went to bed; there was a tide to be got to in the morning.

On the beach at San Lucas there is a war between the pigs and the vultures. Sometimes one side dominates and sometimes the other. On occasion the swine feel a dynamism and demand *Lebensraum,* and in the pride of their species drive the vultures from the decaying offal. And again, when their thousand years of history is over, the vultures spring to arms, tear up treaties, and flap the pigs from the garbage. And on the beach there are certain skinny dogs, without any dynamisms whatever and without racial pride, who nevertheless manage to get the best snacks. They don't thrive on it—always they are meager and skinny and cowardly—but when the *Gauleiter* swine has just captured a fish belly, and before he can shout his second "*Sieg Heil!*" the dog has it.

10

MARCH 18

The tidal series was short. We wished to cover as much ground as possible, to establish as many collecting stations as we could, for we wanted a picture as nearly whole of the Gulf as possible. The next morning we got under way to run the short distance to Pulmo Reef, around the tip and on the eastern shore of the Peninsula. It was a brilliant day, the water riffled and very blue, the sandy beaches of the shore shining with yellow intensity. Above the beaches the low hills were dark with brush. Many people had come to Cape San Lucas, and many had described it. We had read a number of the accounts, and of course agreed with none of them. To a man straight off a yacht, it is a miserable little flea-bitten place, poor and smelly. But to one who puts in hungry, in a storm-beaten boat, it must be a place of great comfort and warmth. These are extremes, but the area in between them also has its multiform conditioning, and what we saw had our conditioning. Once we read a diary, written by a man who came through Panama in 1839. He had read about the place before he got there, but the account he read was about the old city, and in his diary, written after he had gone through, he set down a description of the city he had read about. He didn't know that the town in the book had been destroyed, and that the new one was not even in the same place, but he was not disturbed by these discrepancies. He knew what he would find there and he found it.

There is a curious idea among unscientific men that in scientific writing there is a common plateau of perfectionism. Nothing could be more untrue. The reports of biologists are the measure, not of the science, but of the men themselves. There are as few scientific giants as any other kind. In some reports it is impossible, because of inept expression, to relate the descriptions to the living animals. In some papers collecting places are so mixed or ignored that the animals mentioned cannot be found at all. The same conditioning forces itself into specification as it does into any other kind of observation, and the same faults of carelessness will be found in scientific reports as in the witness chair of a criminal court. It has seemed sometimes that the little men in scientific work assumed the awe-fullness of a priesthood to hide their deficiencies, as the witch-doctor does with his stilts and high masks, as the priesthoods of all cults have, with secret or unfamiliar languages and symbols. It is usually found that only the little stuffy men object to what is called "popularization," by which they mean writing with a clarity understandable to one not familiar with the tricks and codes of the cult. We have not known a single great scientist who could not discourse freely and interestingly with a child. Can it be that the haters of clarity have nothing to say, have observed nothing, have no clear picture of even their own fields? A dull man seems to be a dull man no matter what his field, and of course it is the right of a dull scientist to protect himself with feathers and robes, emblems and degrees, as do other dull men who are potentates and grand imperial rulers of lodges of dull men.

As we neared Pulmo Reef, Tony sent a man up the mast to the crow's-nest to watch for concealed rocks. It is possible to see deep into the water from that high place; the rocks seem to float suddenly up from the bottom like dark shadows. The water in this shallow area was green rather than blue, and the sandy bottom was clearly visible. We pulled in as close as was safe and dropped our anchor. About a mile away we could see the proper reef with the tide beginning to go down on it. On the shore behind the white beach was one of those lonely little *rancherias* we came to know later. Usually a palm or two are planted near by, and by these trees sticking up out of the brush one can locate the houses. There is usually a small corral, a burro or two, a few pigs, and some scrawny chickens. The cattle range wide for food. A dugout canoe lies on the beach, for a good part of the food comes from the sea. Rarely do you see a light from the sea, for the people go to sleep at dusk and awaken with the first light. They must be very lonely people, for they appear on shore the moment a boat anchors, and paddle out in their canoes. At Pulmo Reef the little canoe put off and came alongside. In it were two men and a woman, very ragged, their old clothes patched with the tatters of older clothes. The *serapes* of the men were so thin and threadbare that the light shone through them, and the woman's *rebozo* had long lost its color. They sat in the canoe holding to the side of the *Western Flyer,* and they held their greasy blankets carefully over their noses and mouths to protect themselves from us. So much evil the white man had brought to their ancestors: his breath was poisonous with the lung disease; to sleep with him was to poison the generations. Where he set down his colonies the indigenous people withered and died. He brought industry and trade but no prosperity, riches but no ease. After four hundred years of him these people have ragged clothes and the shame that forces the wearing of them; iron harpoons for their hands, syphilis and tuberculosis; a few of the white man's less complex neuroses, and a curious devotion to a God who was sacrificed long ago in the white man's country. They know the white man is poisonous and they cover their noses against him. They do find us fascinating. However, they sit on the rail for many hours watching us and waiting. When we feed them they eat and are courteous about it, but they did not come for food, they are not beggars. We give the men some shirts and they fold them and put them into the bow of the canoe, but they did not come for clothing. One of the men at last offers us a match-box in which are a few misshapen little pearls like small pale cancers. Five pesos he wants for the pearls, and he knows they aren't worth it. We give him a carton of cigarettes and take his pearls, although we do not want them, for they are ugly little things. Now these three should go, but they do not. They would stay for weeks, not moving nor talking except now and then to one another in soft little voices as gentle as whispers. Their dark eyes never leave us. They ask no questions. They seem actually to be dreaming. Sometimes we asked of the Indians the local names of animals we had taken, and then they consulted together. They seemed to live on remembered things, to be so related to the seashore and the rocky hills and the loneliness that they are these things. To ask about the country is like asking about themselves. "How many toes have you?" "What, toes? Let's see—of course, ten. I have known them all my life, I never thought to count them. Of course it will rain tonight, I don't know why. Something in me tells me I will rain tonight. Of course, I am the whole thing, now that I think about it. I ought to know when I will rain." The dark eyes, whites brown and stained, have curious red lights in the pupils. They seem to be a dreaming people. If finally you must escape their eyes, their timeless dreaming eyes, you have only to say, "*Adiós, señor,*" and they seem to start awake. "*Adiós,*" they say softly. "*Que vaya con Dios.*" And they paddle away. They bring a hush with them, and when they go away one's own voice sounds loud and raw.

We loaded the smaller skiff with collecting materials: the containers and bars, tubes and buckets. We put the Sea-Cow on the stern and it made one of its few mistakes. It thought we were going directly to the beach instead of to the reef a mile away. It started up with a great roar and ran for a quarter of a mile before it became aware of its mistake. It was rarely fooled again. We rowed on to the reef.

Collecting in this region, we always wore rubber boots. There are many animals which sting, some severely, and at least one urchin which is highly poisonous. Some of the worms, such as *Eurythoë,* leave spines in the skin which burn unmercifully. And even a barnacle cut infects readily. It is impossible to wear gloves; one must simply be as careful as possible and look where the finger is going before it is put down. Some of the little beasts are incredibly gallant and ferocious. On one occasion, a moray eel not more than eight inches long lashed out from under a rock, bit one of us on the finger, and retired. If one is not naturally cautious, painful and bandaged hands very soon teach caution. The boots protect one's feet from nearly everything, but there is an urchin which has spines so sharp that they pierce the rubber and break off in the flesh, and they sting badly and usually cause infection.

Pulmo is a coral reef. It has often been remarked that reef-building corals seem to live only on the eastern sides of large land bodies, not on the western sides. This has been noticed many times, and even here at Pulmo the reef-building coral[1] occurs only on the eastern side of the Peninsula. This can have nothing to do with wave-shock or current, but must be governed by another of those unknown factors so ever-present and so haunting to the ecologist.

The complexity of the life-pattern on Pulmo Reef was even greater than at Cape San Lucas. Clinging to the coral, growing on it, burrowing into it, was a teeming fauna. Every piece of the soft material broken off skittered and pulsed with life—little crabs and

[1] *Pocillopora capitata* Verrill.

worms and snails. One small piece of coral might conceal thirty or forty species, and the colors on the reef were electric. The sharp-spined urchins[2] gave us trouble immediately, for several of us, on putting our feet down injudiciously, drove the spines into our toes.

The reef was gradually exposed as the tide went down, and on its flat top the tide pools were beautiful. We collected as widely and rapidly as possible, trying to take a cross-section of the animals we saw. There were purple pendent gorgonians like lacy fans; a number of small spine-covered puffer fish which bloat themselves when they are attacked, erecting the spines; and many starfish, including some purple and gold cushion stars. The club-spined sea-urchins[3] were numerous in their rock niches. They seemed to move about very little, for their niches always just fit them, and have the marks of constant occupation. We took a number of the slim green and brown starfish[4] and the large slim five-rayed starfish with plates bordering the ambulacral grooves.[5] There were numbers of barnacles and several types of brittle-stars. We took one huge, magnificent murex snail. One large hemispherical snail was so camouflaged with little plants, corallines, and other algae that it could not be told from the reef itself until it was turned over. Rock oysters there were, and oysters; limpets and sponges; corals of two types; peanut worms; sea-cucumbers; and many crabs, particularly some disguised in dresses of growing algae which made them look like knobs on the reef until they moved. There were many worms, including our enemy *Eurythoë*, which stings so badly. This worm makes one timid about reaching without looking. The coral clusters were violently inhabited by snapping shrimps, red smooth crabs,[6] and little fuzzy black and white spider crabs.[7] Autotomy in these crabs, shrimps, and brittle-stars is very highly developed. At last, under the reef, we saw a large fleshy gorgonian, or sea-fan, waving gently in the clear water, but it was deep and we could not reach it. One of us took off his clothes and dived for it, expecting at any moment to be attacked by one of those monsters we do not believe

[2] *Arbacia incisa.*

[3] *Eucidaris thouarsii.*

[4] *Phataria unifascialis* Gray.

[5] *Pharia pyramidata.*

[6] *Trapezia* spp.

[7] *Mithrax areolatus.*

in. It was murky under the reef, and the colors of the sponges were more brilliant than in those exposed to greater light. The diver did not stay long; he pulled the large sea-fan free and came up again. And although he went down a number of times, this was the only one of this type of gorgonian he could find. Indeed, it was the only one taken on the entire trip.

The collecting buckets and tubes and jars were very full of specimens—so full that we had constantly to change the water to keep the animals alive. Several large pieces of coral were taken and kept submerged in buckets and later were allowed to lie in stale sea water in one of the pans. This is an interesting thing, for as the water goes stale, the thousands of little roomers which live in the tubes and caves and interstices of the coral come out of hiding and scramble for a new home. Worms and tiny crabs appear from nowhere and are then easily picked up.

The sea bottom inside the reef was of white sand studded with purple and gold cushion stars, of which we collected many. And lying on the sandy bottom were heads and knobs of another coral,[8] much harder and more regularly formed than the reef-building coral. The rush of collecting as much as possible before the tide recovered the reef made us indiscriminate in our collecting, but in the long run this did not matter. For once on board the boat again we could re-collect, going over the pieces of coral and rubble carefully and very often finding animals we had not known were there.

El Pulmo was the only coral reef we found on the entire expedition, and the fauna and even the algae were rather specialized to it. No very great surf could have beaten it, for extremely delicate animals lived on its exposed top where they would have been crushed or washed away had strong seas struck them. And the competition for existence was as great as it had been at San Lucas, but it seemed to us that different methods were employed for frustrating enemies. Whereas at San Lucas speed and ferocity were the attributes of most animals, at Pulmo concealment and camouflage were largely employed. The little crabs wore masks of algae and bryozoa and even hydroids, and most animals had little tunnels or some protected place to run to. The softness of the coral made this possible, where the hard smooth granite of San Lucas had forbidden it. On several occasions we wished for diving equipment, but never

[8] *Porites porosa* Verrill.

more than here at Pulmo, for the under-cut shoreward side of the reef concealed hazy wonders which we could not get at. It is not satisfactory to hold one's breath and to look with unglassed eyes through the dim waters.

The water behind the reef was very warm. We abandoned our boots and, putting on tennis shoes to protect our feet from various stingers, we dived again and again for perfect knobs of coral.

Again we tried to start the Sea-Cow—and then rowed back to the *Western Flyer*. There we complained so bitterly to Tex, the engineer, that he took the evil little thing to pieces. Piece by piece he examined it, with a look of incredulity in his eyes. He admired, I think, the ingenuity which could build such a perfect little engine, and he was astonished at the concept of building a whole motor for the purpose of not running. Having put it together again, he made a discovery. The Sea-Cow would run perfectly out of water—that is, in a barrel of water with the propeller and cooling inlet submerged. Placed thus, the Sea-Cow functioned perfectly and got good mileage.

Immediately on arriving back at the *Western Flyer* we pulled up the anchor and got under way again. It was efficient that we preserve and label while we sailed as long as the sea was calm, and now it was very calm. The great collection from the reef required every enameled pan and glass dish we had. The killing and relaxing and preserving took us until dark, and even after dark we sat and made the labels to go into the tubes. As the jars filled and were labeled, we put them back in their corrugated-paper cartons and stowed them in the hold. The corked tubes were tested for leaks, then wrapped in paper toweling and stacked in boxes. Thus there was very small loss from breakage or leakage, and by labeling the same day as collecting, there had thus far been virtually no confusion in the tabulation of animals. But we knew already that we had made one error in planning: we had not brought nearly enough small containers. It is best to place an animal alone in a jar or a tube which accommodates him, but not too freely. The enormous numbers of animals we took strained our resources and containers long before we were through.

As we moved up the Gulf, the mirage we had heard about began to distort the land. While it is worse on the Sonora coast, it is sufficiently interesting on the Peninsula to produce a heady, crazy feeling in the observer. As you pass a headland it suddenly splits off and becomes an island and then the water seems to stretch inward and pinch it to a mushroom-shaped cliff, and finally to liberate it from the earth entirely so that it hangs in the air over the water. Even a short distance offshore one cannot tell what the land really looks like. Islands too far off, according to the map, are visible; while others which should be near by cannot be seen at all until suddenly they come bursting out of the mirage. The whole surrounding land is unsubstantial and changing. One remembers the old stories of invisible kingdoms where princes lived with ladies and dragons for company; and the more modern fairy-tales in which heroes drift in and out of dimensions more complex than the original three. We are open enough to miracles of course, but what must have been the feeling of the discovering Spaniards? Miracles were daily happenings to them. Perhaps to that extent their feet were more firmly planted on the ground. Subject as they were to the constant apparitions of saints, to the trooping of holy virgins into their dreams and reveries, perhaps mirages were commonplaces. We have seen many miraculous figures in Mexico. They are usually Christs which have supernaturally appeared on mountains or in caves and usually at times of crisis. But it does seem odd that the heavenly authorities, when they wished a miraculous image to appear, invariably chose bad Spanish wood-carving of the seventeenth century. But perhaps art criticism in heaven was very closely related to the sensibilities of the time. Certainly it would have been a little shocking to find an Epstein Christ under a tree on a mountain in Mexico, or a Brancusi bird, or a Dali *Descent from the Cross.*

It must have been a difficult task for those first sturdy Jesuit fathers to impress the Indians of the Gulf. The very air here is miraculous, and outlines of reality change with the moment. The sky sucks up the land and disgorges it. A dream hangs over the whole region, a brooding kind of hallucination. Perhaps only the shock of seventeenth-century wood-carving could do the trick; surely the miracle must have been very virile to be effective.

Tony grew restive when the mirage was working, for here right and wrong fought before his very eyes, and how could one tell which was error? It is very well to say, "The land is here and what blots it out is a curious illusion caused by light and air and moisture," but if one is steering a boat, he must sail by what he sees, and if air and light and moisture—three realities—plot together and perpetrate a lie, what is a realistic man to believe? Tony did not like the mirage at all.

While we worked at the specimens, the trolling lines were out and we caught another skipjack, large and fat and fast. As it came in on the line, one of us ran for the moving-picture camera, for we wanted to record on color film the changing tints and patterns of the fish's dying. But the exposure was wrong as usual, and we did not get it.

Near the moving boat swordfishes played about. They seemed to play in pure joy or exhibitionism. It is thought that they leap to clear themselves of parasites; they jump clear of the water and come crashing down, and sometimes they turn over in the air and flash in the sunshine. This afternoon, too, we saw the first specimens of the great manta ray (a giant skate), and we rigged the harpoons and coiled the line ready. One light harpoon just pierced a swordfish's tail, but he swished away, for the barb had not penetrated. And we did not turn and pursue the great rays, for we wished to anchor that night near Point Lobos on Espíritu Santo Island.

In the evening we came near to it, but as we prepared to anchor, the wind sprang up full on us, and Tony decided to run for the shelter of Pescadero Point on the mainland. The wind seemed to grow instantly out of the evening, and the sea with it. The jars and collecting pans were in danger of flying overboard. For half an hour we were very busy tying the equipment down and removing the flapping canvas we had stretched to keep the sun off our specimen pans. Under the powerful wind we crossed the channel which leads to La Paz, and saw the channel light—the first one we had seen since the big one on the false cape. This one seemed very strange in the Gulf. The waves were not high, but the wind blew with great intensity, making whitecaps rather than rollers, and only when we ran in under Pescadero Point did we drop the wind. We eased in slowly, sounding as we went. When the anchor was finally down we cooked and ate the skipjack, a most delicious fish. And after dinner a group action took place.

We carried no cook and dishwasher; it had been understood that we would all help. But for some time Tex had been secretly mutinous about washing dishes. At the proper times he had things to do in the engine-room. He might have succeeded in this crime, if he had ever varied his routine, but gradually a suspicion grew on us that Tex did not like to wash dishes. He denied this vigorously. He said he liked very much to wash dishes. He appealed to our reason. How would we like it, he argued, if we were forever in the engine-room, getting our hands dirty? There was danger down there too, he said. Men had been killed by engines. He was not willing to see us take the risk. We met his arguments with a silence that made him nervous. He protested then that he had once washed dishes from west Texas to San Diego without stopping, and that he had learned to love it so much that he didn't want to be selfish about it now. A circle of cold eyes surrounded him. He began to sweat. He said that later (he didn't say how much later) he was going to ask us for the privilege of washing all the dishes, but right now he had a little job to do in the engine-room. It was for the safety of the ship, he said. No one answered him. Then he cried, "My God, are you going to hang me?" At last Sparky spoke up, not unkindly, but inexorably. "Tex," he said, "you're going to wash 'em or you're going to sleep with 'em." Tex said, "Now just as soon as I do one little job there's nothing I'd rather do than wash four or five thousand dishes." Each of us picked up a load of dishes, carried them in, and laid them gently in Tex's bunk. He got up resignedly then and carried them back and washed them. He didn't grumble, but he was broken. Some joyous light had gone out of him, and he never did get the catsup out of his blankets.

That night Sparky worked at the radio and made contact with the fishing fleet that was operating in the region from Cedros Island and around the tip into the Gulf, fishing for tuna. Fishermen are no happier than farmers. It is difficult to see why anyone becomes a farmer or a fisherman. Dreadful things happen to them constantly: they lose their nets; the fish are wild; sea-lions get into the nets and tear their way out; snags are caught; there are no fish, and the price high; there are too many fish, and the price is low; and if some means could be devised so that the fish swam up to a boat, wriggled up a trough, squirmed their way into the fish-hold, and pulled ice over themselves with their own fins, the imprecations would be terrible because they had not removed their own entrails and brought their own ice. There is no happiness for fishermen anywhere. Cries of anguish at the injustice of the elements inundated the short-wave receiver as we lay at anchor.

The pattern of a book, or a day, of a trip, becomes a characteristic design. The factors in a trip by boat, the many-formed personality phases all shuffled together, changing a little to fit into the box and yet bringing their own lumps and corners, make the trip. And from all these factors your expedition has a character of its own, so that one may say of it, "That was a good, kind trip." Or, "That was a mean one." The character of the whole becomes defined and definite. We ran from collecting station to new collecting station, and when the night came and the anchor was dropped, a quiet came over the boat and the trip slept. And then we talked and speculated, talked and drank beer. And our discussions ranged from the loveliness of remembered women to the complexities of relationships in every other field. It is very easy to grow tired at collecting; the period of a low tide is about all men can endure. At first the rocks are bright and every moving animal makes his mark on the attention. The picture is wide and colored and beautiful. But after an hour and a half the attention centers weary, the colors fade, and the field is likely to narrow to an individual animal. Here one may observe his own world narrowed down until interest and, with it, observation, flicker and go out. And what if with age this weariness become permanent and observation dim out and not recover? Can this be what happens to so many men of science? Enthusiasm, interest, sharpness, dulled with a weariness until finally they retire into easy didacticism? With this weariness, this stultification of the attention centers, perhaps there comes the pained and sad memory of what the old excitement was like, and regret might turn to envy of the men who still have it. Then out of the shell of didacticism, such a used-up man might attack the unwearied, and he would have in his hands proper weapons of attack. It does seem certain that to a wearied man an error in a mass of correct data wipes out all the correctness and is a focus for attack; whereas the unwearied man, in his energy and receptivity, might consider the little dross of error a by-product of his effort. These two may balance and produce a purer thing than either in the end. These two may be the stresses which hold up the structure, but it is a sad thing to see the interest in interested men thin out and weaken and die. We have known so many professors who once carried their listeners high on their single enthusiasm, and have seen these same men finally settle back comfortably into lectures prepared years before and never vary them again. Perhaps this is the same narrowing we observe in relation to ourselves and the tide pool—a man looking at reality brings his own limitations to the world. If he has strength and energy of mind the tide pool stretches both ways, digs back to electrons and leaps space into the universe and fights out of the moment into non-conceptual time. Then ecology has a synonym which is ALL.

It is strange how the time sense changes with different peoples. The Indians who sat on the rail of the *Western Flyer* had a different time sense—"time-world" would be the better term—from ours. And we think we can never get into them unless we can invade that time-world, for this expanding time seems to trail an expanding universe, or perhaps to lead it. One considers the durations indicated in geology, in paleontology, and, thinking out of our time-world with its duration between time-stone and time-stone, says, "What an incredible interval!" Then, when one struggles to build some picture of astro-physical time, he is faced with a light-year, a thought-deranging duration unless the relativity of all things intervenes and time expands and contracts, matching itself relatively to the pulsings of a relative universe.

It is amazing how the strictures of the old teleologies infect our observation, causal thinking warped by hope. It was said earlier that hope is a diagnostic human trait, and this simple cortex symptom seems to be a prime factor in our inspection of our universe. For hope implies a change from a present bad condition to a future better one. The slave hopes for freedom, the weary man for rest, the hungry for food. And the feeders of hope, economic and religious, have from these simple strivings of dissatisfaction managed to create a world picture which is very hard to escape. Man grows toward perfection; animals grow toward man; bad grows toward good; and down toward up, until our little mechanism, hope, achieved in ourselves probably to cushion the shock of thought, manages to warp our whole world. Probably when our species developed the trick of memory and with it the counterbalancing projection called "the future," this shock-absorber, hope, had to be included in the series, else the species would have destroyed itself in despair. For if ever any man were deeply and unconsciously sure that his future would be no better than his past, he might deeply wish to cease to live. And out of this therapeutic poultice we build our iron teleologies and twist the tide pools and the stars into the pattern. To most men the most hateful statement possible is, "*A thing is because it is.*" Even those who have managed to drop the leading-strings of a Sunday-school deity are still led by the unconscious teleology of their developed trick. And in saying that hope cushions the shock of experience, that one trait balances the directionalism of another, a teleology is implied, unless one know or feel or think that we *are* here, and that without this balance, hope, our species in its blind mutation might have joined many, many others in extinction. Dr. Torsten Gislén, in his fine paper on fossil echinoderms called "Evolutional Series toward Death and Renewal,"[9] has shown that as often as not, in his studied group at least, mutations have had destructive, rather than survival value. Extending this thesis, it is interesting to think of the mutations of our own species. It is said and thought there has been none in historical times. We wonder, though, where in man a mutation might take place. Man is the only animal whose interest and whose drive are outside himself. Other animals may dig holes to live in; may weave nests or take possession of hollow trees. Some species, like bees or spiders, even create complicated homes, but they do it with the fluids and processes of their own bodies. They make little impression on the world. But the world is furrowed and cut, torn and blasted by man. Its flora has been swept away and changed; its mountains torn down by man; its flat lands littered by the debris of his living. And these changes have been wrought, not because any inherent technical ability has demanded them, but because his desire has created that technical ability. Physiological man does not require this paraphernalia to exist, but the whole man does. He is the only animal who lives outside of himself, whose drive is in external things—property, houses, money, concepts of power. He lives in his cities and his factories, in his business and job and art. But having projected himself into these external complexities, he *is* them. His house, his automobile are a part of him and a large part of him. This is beautifully demonstrated by a thing doctors know —that when a man loses his possessions a very common result is sexual impotence. If then the projection, the preoccupation of man, lies in external things so that even his subjectivity is a mirror of houses and cars and grain elevators, the place to look for his mutation would be in the direction of his drive, or in other words in the external things he deals with. And here we can indeed readily

[9] *Ark. f. zool. K. Svenska Vetens.*, Vol. 26 A, No. 16, Stockholm, Jan. 1934.

find evidence of mutation. The industrial revolution would then be indeed a true mutation, and the present tendency toward collectivism, whether attributed to Marx or Hitler or Henry Ford, might be as definite a mutation of the species as the lengthening neck of the evolving giraffe. For it must be that mutations take place in the direction of a species drive or preoccupation. If then this tendency toward collectivization is mutation there is no reason to suppose it is for the better. It is a rule in paleontology that ornamentation and complication precede extinction. And our mutation, of which the assembly line, the collective farm, the mechanized army, and the mass production of food are evidences or even symptoms, might well correspond to the thickening armor of the great reptiles—a tendency that can end only in extinction. If this should happen to be true, nothing stemming from thought can interfere with it or bend it. Conscious thought seems to have little effect on the action or direction of our species. There is a war now which no one wants to fight, in which no one can see a gain—a zombie war of sleep-walkers which nevertheless goes on out of all control of intelligence. Some time ago a Congress of honest men refused an appropriation of several hundreds of millions of dollars to feed our people. They said, and meant it, that the economic structure of the country would collapse under the pressure of such expenditure. And now the same men, just as honestly, are devoting many billions to the manufacture, transportation, and detonation of explosives to protect the people they would not feed. And it must go on. Perhaps it is all a part of the process of mutation and perhaps the mutation will see us done for. We have made our mark on the world, but we have really done nothing that the trees and creeping plants, ice and erosion, cannot remove in a fairly short time. And it is strange and sad and again symptomatic that most people, reading this speculation which is *only* speculation, will feel that it is a treason to our species so to speculate. For in spite of overwhelming evidence to the contrary, the trait of hope still controls the future, and man, not a species, but a triumphant race, will approach perfection, and, finally, tearing himself free, will march up the stars and take his place where, because of his power and virtue, he belongs: on the right hand of the $\sqrt[\pi]{-1}$. From which majestic seat he will direct with pure intelligence the ordering of the universe. And perhaps when that occurs—when our species progresses toward extinction or marches into the forehead of God—there will be certain degenerate groups left behind, say, the Indians of Lower California, in the shadows of the rocks or sitting motionless in the dugout canoes. They may remain to sun themselves, to eat and starve and sleep and reproduce. Now they have many legends as hazy and magical as the mirage. Perhaps then they will have another concerning a great and godlike race that flew away in four-motored bombers to the accompaniment of exploding bombs, the voice of God calling them home.

Nights at anchor in the Gulf are quiet and strange. The water is smooth, almost solid, and the dew is so heavy that the decks are soaked. The little waves rasp on the shell beaches with a hissing sound, and all about in the darkness the fishes jump and splash. Sometimes a great ray leaps clear and falls back on the water with a sharp report. And again, a school of tiny fishes whisper along the surface, each one, as it breaks clear, making the tiniest whisking sound. And there is no feeling, no smell, no vibration of people in the Gulf. Whatever it is that makes one aware that men are about is not there. Thus, in spite of the noises of waves and fishes, one has a feeling of deadness and of quietness. At anchor, with the motor stopped, it is not easy to sleep, and every little sound starts one awake. The crew is restless and a little nervous. If a dog barks on shore or a cow bellows, we are reassured. But in many places of anchorage there were utterly no sounds associated with man. The crew read books they have not known about—Tony reads *Studs Lonigan* and says he does not like to see such words in print. And we are reminded that we once did not like to hear them spoken because we were not used to them. When we became used to hearing them, they took their place with the simple speech-sounds of the race of man. Tony read on in *Studs Lonigan,* and the shock of the new words he had not seen printed left him and he grew into the experience of Studs. Tiny read the book too. He said, "It's like something that happened to me."

Sometimes in the night a little breeze springs up and the boat tugs experimentally at the anchor and swings slowly around. There is nothing so quiet as a boat when the motor has stopped; it seems to lie with held breath. One gets to longing for the deep beat of the cylinders.

11

MARCH 20

We had marked the southern end of Espíritu Santo Island as our next collecting stop. This is a long narrow island which makes the northern side of the San Lorenzo Channel. It is mountainous and stands high and sheer from the blue water. We wanted particularly to collect there so that we could contrast the fauna of the eastern tip of this island with that of the secluded and protected bay of La Paz. Throughout we attempted to work in stations in the same area which nevertheless contrasted conditions for living, such as wave-shock, bottom, rock formation, exposure, depth, and so forth. The most radical differences in life forms are discovered in this way.

Early in the morning we sailed from our shelter under Pescadero Point and crossed the channel again. It was a very short run. There were many manta rays cruising slowly near the surface, with only the tips of their "wings" protruding above the water. They seemed to hover, and when we approached too near, they disappeared into the blue depths. Their effortless speed is astonishing. On the lines we caught two yellowfin tunas,[1] speedy and efficient fish. They struck the line so hard that it is impossible to see why they did not tear their heads off.

We anchored near a bouldery shore. This would be the first station in the Gulf where we would be able to turn over rocks, and a new ecological set-up was indicated by the fact that the small boulders rested in sand.

This time everyone but Tony went ashore. Sparky and Tiny were already developing into good collectors, and now Tex joined us and quickly became excited in the collecting. We welcomed this help, for in general work, what with the shortness of the time and the large areas to be covered, the more hands and eyes involved, the better. Besides, these men who lived by the sea had a great respect for the sea and all its inhabitants. Association with the sea does not breed contempt.

[1] *Neothunnus macropterus.*

The boulders on this beach were almost a perfect turning-over size—heavy enough to protect the animals under them from grinding by the waves, and light enough to be lifted. They were well coated with short algae and bedded in very coarse sand. The dominant species on this beach was a sulphury cucumber,[2] a dark, almost black-green holothurian which looks as though it were dusted with sulphur. As the tide dropped on the shallow beach we saw literally millions of these cucumbers. They lay in clusters and piles between the rocks and under the rocks, and as the tide went down and the tropical sun beat on the beach, many of them became quite dry without apparent injury. Most of these holothurians were from five to eight inches long, but there were great numbers of babies, some not more than an inch in length. We took a great many of them.

Easily the second most important animal of this shore in point of quantity was the brittle-star. We had read of their numbers in the Gulf and here they were, mats and clusters of them, giants under the rocks. It was simple to pick up a hundred at a time in black, twisting, squirming knots. There were five species of them, and these we took in large numbers also, for in preservation they sometimes cast off their legs or curl up into knots, and we wished to have a number of perfect specimens. Starfish were abundant here and we took six varieties. The difference between the brittle-star and the starfish is interestingly reflected in the scientific names—"Ophio" is a Greek root signifying "serpent"—the round compact body and long serpent-like arms of the brittle-star are suggested in the generic name "ophiuran," while the more truly star-like form of the starfish is recognizable in the Greek root "aster," which occurs in so many of its proper names, "Heliaster," "Astrometis," etc. We found three species of urchins, among them the very sharp-spined and poisonous *Centrechinus mexicanus;* approximately ten different kinds of crabs, four of shrimps, a number of anemones of various types, a great number of worms, including our enemy *Eurythoë,* which seems to occur everywhere in the Gulf, several species of naked mollusks, and a good number of peanut worms. The rocks and the sand underneath them were heavily populated. There were chitons and keyhole limpets, a number of species of clams, flatworms, sponges, bryozoa, and numerous snails.

[2] *Holothuria lubrica.*

Again the collecting buckets were very full, but already we had begun the elimination of animals to be taken. On this day we took enough of the sulphury cucumbers and brittle-stars for our needs. These were carefully preserved, but when found again at a new station they would simply be noted in the collecting record, unless some other circumstance such as color change or size variation prevailed. Thus, as we proceeded, we gradually stopped collecting certain species and only noted them as occurring.

On board the *Western Flyer*, again we laid out the animals in pans and prepared them for anesthetization. In one of the sea-cucumbers we found a small commensal fish[3] which lived well inside the anus. It moved in and out with great ease and speed, resting invariably head inward. In the pan we ejected this fish by a light pressure on the body of the cucumber, but it quickly returned and entered the anus again. The pale, colorless appearance of this fish seemed to indicate that it habitually lived there.

It is interesting to see how areas are sometimes dominated by one or two species. On this beach the yellow-green cucumber was everywhere, with giant brittle-stars a close second. Neither of these animals has any effective offensive property as far as we know, although neither of them seems to be a delicacy enjoyed by other animals. There does seem to be a balance which, when passed by a certain species, allows that animal numerically to dominate a given area. When this threshold of successful reproduction and survival is crossed, the area becomes the special residence of this form. Then it seems other animals which might be either hostile or perhaps the prey of the dominating animal would be wiped out or would desert the given area. In many cases the arrival and success of a species seem to be by chance entirely. In some northern areas, where the ice of winter yearly scours and cleans the rocks, it has been noted that summer brings sometimes one dominant species and sometimes another, the success factor seeming to be prior arrival and an early start.[4] With marine fauna, as with humans, priority and possession appear to be vastly important to survival and dominance. But sometimes it is found that the very success of an animal is its downfall.

[3] *Encheliophiops hancocki* Reid.

[4] Gislén, T., "Epibioses of the Gullmar Fjord II." 1930, p. 157. Kristinebergs Zool. Sta. 1877–1927, *Skrift. ut. av K. Svenska Vetens.* N:r 4.

There are examples where the available food supply is so exhausted by the rapid and successful reproduction that the animal must migrate or die. Sometimes, also, the very by-products of the animals' own bodies prove poisonous to a too great concentration of their own species.

It is difficult, when watching the little beasts, not to trace human parallels. The greatest danger to a speculative biologist is analogy. It is a pitfall to be avoided—the industry of the bee, the economics of the ant, the villainy of the snake, all in human terms have given us profound misconceptions of the animals. But parallels are amusing if they are not taken too seriously as regards the animal in question, and are downright valuable as regards humans. The routine of changing domination is a case in point. One can think of the attached and dominant human who has captured the place, the property, and the security. He dominates his area. To protect it, he has police who know him and who are dependent on him for a living. He is protected by good clothing, good houses, and good food. He is protected even against illness. One would say that he is safe, that he would have many children, and that his seed would in a short time litter the world. But in his fight for dominance he has pushed out others of his species who were not so fit to dominate, and perhaps these have become wanderers, improperly clothed, ill fed, having no security and no fixed base. These should really perish, but the reverse seems true. The dominant human, in his security, grows soft and fearful. He spends a great part of his time in protecting himself. Far from reproducing rapidly, he has fewer children, and the ones he does have are ill protected inside themselves because so thoroughly protected from without. The lean and hungry grow strong, and the strongest of them are selected out. Having nothing to lose and all to gain, these selected hungry and rapacious ones develop attack rather than defense techniques, and become strong in them, so that one day the dominant man is eliminated and the strong and hungry wanderer takes his place.

And the routine is repeated. The new dominant entrenches himself and then softens. The turnover of dominant human families is very rapid, a few generations usually sufficing for their rise and flowering and decay. Sometimes, as in the case of Hearst, the rise and glory and decay take place in one generation and nothing is left. One dominant thing sometimes does survive and that is not even well defined; some quality of the spirit of an individual continues to dominate. Whereas the great force which was Hearst has died before the death of the man and will soon be forgotten except perhaps as a ridiculous and vulgar fable, the spirit and thought of Socrates not only survive, but continue as living entities.

There is a strange duality in the human which makes for an ethical paradox. We have definitions of good qualities and of bad; not changing things, but generally considered good and bad throughout the ages and throughout the species. Of the good, we think always of wisdom, tolerance, kindliness, generosity, humility; and the qualities of cruelty, greed, self-interest, graspingness, and rapacity are universally considered undesirable. And yet in our structure of society, the so-called and considered good qualities are invariable concomitants of failure, while the bad ones are the cornerstones of success. A man—a viewing-point man—while he will love the abstract good qualities and detest the abstract bad, will nevertheless envy and admire the person who through possessing the bad qualities has succeeded economically and socially, and will hold in contempt that person whose good qualities have caused failure. When such a viewing-point man thinks of Jesus or St. Augustine or Socrates he regards them with love because they are the symbols of the good he admires, and he hates the symbols of the bad. But actually he would rather be successful than good. In an animal other than man we would replace the term "good" with "weak survival quotient" and the term "bad" with "strong survival quotient." Thus, man in his thinking or reverie status admires the progression toward extinction, but in the unthinking stimulus which really activates him he tends toward survival. Perhaps no other animal is so torn between alternatives. Man might be described fairly adequately, if simply, as a two-legged paradox. He has never become accustomed to the tragic miracle of consciousness. Perhaps, as has been suggested, his species is not set, has not jelled, but is still in a state of becoming, bound by his physical memories to a past of struggle and survival, limited in his futures by the uneasiness of thought and consciousness.

Back on the *Western Flyer,* Sparky cooked the tuna in a sauce of tomatoes and onions and spices and we ate magnificently. Each rock turned over had not been heavy, but we had turned over many tons of rocks in all. And now the work with the animals had to go on, the preservation and labeling. But we rested and drank a little beer, which in this condition of weariness is rest itself.

While we were eating, a boat came alongside and two Indians climbed aboard. Their clothing was better than that of the poor people of the day before. They were, after all, within a day's canoe trip of La Paz, and some of the veneer of that city had stuck to them. Their clothing was patched and ragged, but at least not falling apart from decay. We asked Sparky and Tiny to bring them a little wine, and after two glasses they became very affable, making us think of the intolerance of the Indian for alcohol. Later it developed that Sparky and Tiny had generously laced the wine with whisky, which proved just the opposite about the Indians' tolerance for alcohol. None of us could have drunk two tumblers, half whisky and half wine, but these men did and became gay and companionable. They were barefoot and carried the iron harpoons of the region, and in the bottom of their canoe lay a huge fish. Their canoe was typical of the region and was interesting. There are no large trees in the southern part of the Peninsula, hence all the canoes come from the mainland, most of them being made near Mazatlán. They are double-ended canoes carved from a single log of light wood, braced inside with struts. Sometimes a small sail is set, but ordinarily they are paddled swiftly by two men, one at either end. They are seaworthy and fast. The wood inside and out is covered with a thin layer of white or blue plaster, waterproof and very hard. This is made by the people themselves and applied regularly. It is not a paint, but a hard, shell-like plaster, and we could not learn how it is made although this is probably well known to many people. Equipped with one of these canoes, an iron harpoon, a pair of trousers, shirt, and hat, a young man is fairly well set up in life. In fact, the acquiring of a Nayarit canoe will probably give a young man so much security in his own eyes and make him so desirable in the eyes of others that he will promptly get married.

It is said so often and in such ignorance that Mexicans are contented, happy people. "They don't want anything." This, of course, is not a description of the happiness of Mexicans, but of the unhappiness of the person who says it. For Americans, and probably all northern peoples, are all masses of wants growing out of inner insecurity. The great drive of our people stems from insecurity. It is often considered that the violent interest in little games, the mental rat-mazes of contract bridge, and the purposeful striking of little white balls with sticks, comes from an inner sterility. But more likely it comes from an inner complication. Boredom arises not so often from too little to think about, as from too much, and none of it clear nor clean nor simple. Bridge is a means of forgetting the thousands of little irritations of a mind over-crowded with anarchy. For bridge has a purpose, that of taking as many tricks as possible. The end is clear and very simple. But nothing in the lives of bridge-players is clean-cut, and no ends are defined. And so they retire into some orderly process, even in a game, from the messy complication of their lives. It is possible, although we do not know this, that the poor Mexican Indian is a little less messy in his living, having a baby, spearing a fish, getting drunk, backing a political candidate; each one of these is a clear, free process, ending in a result. We have thought of this in regard to the bribes one sometimes gives to Mexican officials. This is universally condemned by Americans, and yet it is a simple, easy process. A bargain is struck, a price named, the money paid, a graceful compliment exchanged, the service performed, and it is over. He is not your man nor you his. A little process has been terminated. It is rather like the old-fashioned buying and selling for cash or produce.

We find we like this cash-and-carry bribery as contrasted with our own system of credits. With us, no bargain is struck, no price named, nothing is clear. We go to a friend who knows a judge. The friend goes to the judge. The judge knows a senator who has the ear of the awarder of contracts. And eventually we sell five carloads of lumber. But the process has only begun. Every member of the chain is tied to every other. Ten years later the son of the awarder of contracts must be appointed to Annapolis. The senator must have traffic tickets fixed for many years to come. The judge has a political lien on your friend, and your friend taxes you indefinitely with friends who need jobs. It would be simpler and cheaper to go to the awarder of contracts, give him one-quarter of the price of the lumber, and get it over with. But that is dishonest, that is a bribe. Everyone in the credit chain eventually hates and fears everyone else. But the bribe-bargain, having no enforcing mechanism, promotes mutual respect and a genuine liking. If the accepter of a bribe cheats you, you will not go to him again and he will soon have to leave the public service. But if he fulfills his contract, you have a new friend whom you can trust.

We do not know whether Mexicans are happier than we; it is probable that they are exactly as happy. However, we do know that the channels of their happiness or unhappiness are different from ours, just as their time sense is different. We can invade neither, but it is some gain simply to know that it is so.

As the men on our deck continued with what we thought was wine and they probably considered some expensive foreign beverage (it must have tasted bad enough to be very foreign and very expensive), they uncovered a talent for speech we have often noticed in these people. They are natural orators, filling their sentences with graceful forms, with similes and elegant parallels. Our oldest man delivered for us a beautiful political speech. He was an ardent admirer of General Almazán, who was then a candidate for the Mexican presidency. Our Indian likened the General militarily to the god of war, but whether Mars, or Huitzilopochtli, he did not say. In physical beauty the General stemmed from Apollo, not he of the Belvedere, but an earlier, sturdier Apollo. In kindness and forethought and wisdom Almazán was rather above the lesser deities. Our man even touched on the General's abilities in bed, although how he knew he did not say. We gathered, though, that the General was known and well thought of in this respect by his total feminine constituency. "He is a strong man," said our orator, holding himself firmly upright by the port stay. One of us interposed, "When he is elected there will be more fish in the sea for the poor people of Mexico." And the orator nodded wisely. "That is so my friend," he said. It was later that we learned that General Camacho, the other candidate, had many of the same beautiful qualities as General Almazán. And since he won, perhaps he had them more highly developed. For political virtues always triumph, and when two such colossi as these oppose each other, one can judge their relative excellences only by counting the vote.

We had known that sooner or later we must develop an explanation for what we were doing which would be short and convincing. It couldn't be the truth because that wouldn't be convincing at all. How can you say to a people who are preoccupied with getting enough food and enough children that you have come to pick up useless little animals so that perhaps your world picture will be enlarged? That didn't even convince us. But there had to be a story, for everyone asked us. One of us had once taken a long walking trip through the southern United States. At first he had tried to explain that he did it because he liked to walk and because he saw and felt the country better that way. When he gave this explanation there was unbelief and dislike for him. It sounded like a lie. Finally a man said to him, "You can't fool me, you're doing it on a bet." And after that, he used this explanation, and everyone liked and understood him from then on. So with these men we developed our story and stuck to it thereafter. We were collecting curios, we said. These beautiful little animals and shells, while they abounded so greatly here as to be valueless, had, because of their scarcity in the United States, a certain value. They would not make us rich but it was at least profitable to take them. And besides, we liked taking them. Once we had developed this story we never had any more trouble. They all understood us then, and brought us what they thought were rare articles for the collection. They considered that we might get very rich. Thank heaven they do not know that when at last we came back to San Diego the customs fixed a value on our thousands of pickled animals of five dollars. We hope these Indians never find it out; we would go down steeply in their estimations.

Our men went away finally a trifle intoxicated, but not forgetting to take an armload of empty tomato cans. They value tin cans very highly.

It would not have done to sail for La Paz harbor that night, for the pilot has short hours and any boat calling for him out of his regular hours must pay double. But we wanted very much to get to La Paz; we were out of beer and already the water in our tanks was stale-tasting. It had seemed to us that it was stale when we put it in and time did not improve it. It isn't likely that we would have died of thirst. The second or third day would undoubtedly have seen us drinking the unpleasant stuff. But there were other reasons why we longed for La Paz. Cape San Lucas had not really been a town, and our crew had convinced itself that it had been a very long time out of touch with civilization. In civilization we think they included some items which, if anything, are attenuated in highly civilized groups. In addition, there is the genuine fascination of the city of La Paz. Everyone in the area knows the greatness of La Paz. You can get anything in the world there, they say. It is a huge place—not of course so monstrous as Guaymas or Mazatlán, but beautiful out of all comparison. The Indians paddle hundreds of miles to be at La Paz on a feast day. It is a proud thing to have been born in La Paz, and a cloud of delight hangs over the distant city from the time when it was the great pearl center of the world. The robes of the Spanish kings and the stoles of bishops in Rome were stiff with the pearls from La Paz. There's a magic-carpet sound to the name, anyway. And it is an old city, as cities in the West are old, and very venerable in the eyes of Indians of the Gulf. Guaymas is busier, they say, and Mazatlán gayer, perhaps, but La Paz is *antigua*.

The Gulf and Gulf ports have always been unfriendly to colonization. Again and again attempts were made before a settlement would stick. Humans are not much wanted on the Peninsula. But at La Paz the pearl oysters drew men from all over the world. And, as in all concentrations of natural wealth, the terrors of greed were let loose on the city again and again. An event which happened at La Paz in recent years is typical of such places. An Indian boy by accident found a pearl of great size, an unbelievable pearl. He knew its value was so great that he need never work again. In his one pearl he had the ability to be drunk as long as he wished, to marry any one of a number of girls, and to make many more a little happy too. In his great pearl lay salvation, for he could in advance purchase masses sufficient to pop him out of Purgatory like a squeezed watermelon seed. In addition he could shift a number of dead relatives a little nearer to Paradise. He went to La Paz with his pearl in his hand and his future clear into eternity in his heart. He took his pearl to a broker and was offered so little that he grew angry, for he knew he was cheated. Then he carried his pearl to another broker and was offered the same amount. After a few more visits he came to know that the brokers were only the many hands of one head and that he could not sell his pearl for more. He took it to the beach and hid it under a stone, and that night he was clubbed into unconsciousness and his clothing was searched. The next night he slept at the house of a friend and his friend and he were injured and bound and the whole house searched. Then he went inland to lose his pursuers and he was waylaid and tortured. But he was very angry now and he knew what he must do. Hurt as he was he crept back to La Paz in the night and he skulked like a hunted fox to the beach and took out his pearl from under the stone. Then he cursed it and threw it as far as he could into the channel. He was a free man again with his soul in danger and his food and shelter insecure. And he laughed a great deal about it.

This seems to be a true story, but it is so much like a parable that it almost can't be. This Indian boy is too heroic, too wise. He knows too much and acts on his knowledge. In every way, he goes contrary to human direction. The story is probably true, but we don't believe it; it is far too reasonable to be true.

La Paz, the great city, was only a little way from us now, we could almost see its towers and smell its perfume. And it was right that it should be so hidden here out of the world, inaccessible except to the galleons of a small boy's imagination.

While we were anchored at Espíritu Santo Island a black yacht went by swiftly, and on her awninged after-deck ladies and gentlemen in white clothing sat comfortably. We saw they had tall cool drinks beside them and we hated them a little, for we were out of beer. And Tiny said fiercely, "Nobody but a pansy'd sail on a thing like that." And then more gently, "But I've never been sure I ain't queer." The yacht went down over the horizon, and up over the horizon climbed an old horror of a cargo ship, dirty and staggering. And she stumbled on toward the channel of La Paz; her pumps must have been going wide open. Later, at La Paz, we saw her very low in the water in the channel. We said to a man on the beach, "She is sinking." And he replied calmly, "She always sinks."

On the *Western Flyer,* vanity had set in. Clothing was washed unmercifully. The white tops of caps were laundered, and jeans washed and patted smooth while wet and hung from the stays to dry. Shoes were even polished and the shaving and bathing were deafening. The sweet smell of unguents and hair oils, of deodorants and lotions, filled the air. Hair was cut and combed; the mirror over the washstand behind the deckhouse was in constant use. We regarded ourselves in the mirror with the long contemplative coy looks of chorus girls about to go on stage. What we found was not good, but it was the best we had. Heaven knows what we expected to find in La Paz, but we wanted to be beautiful for it.

And in the morning, when we got under way, we washed the fish blood off the decks and put away the equipment. We coiled the lines in lovely spirals and washed all the dishes. It seemed to us we made a rather gallant show, and we hoped that no beautiful yacht was anchored in La Paz. If there were a yacht, we would be tough and seafaring, but if no such contrast was available some of us at least proposed to be not a little jaunty. Even the least naïve of us expected Spanish ladies in high combs and mantillas to be promenading along the beach. It would be rather like the opening scene of a Hollywood production of *Life in Latin America*, with dancers in the foreground and cabaret tables upstage from which would rise a male chorus to sing "I met my love in La Paz—satin and Latin she was."

We assembled on top of the deckhouse, the *Coast Pilot* open in front of us. Even Tony had succumbed; he wore a gaudy white seaman's cap with a gold ornament on the front of it which seemed to be a combination of field artillery and submarine service, except that it had an arrow-pierced heart superimposed on it.

We have so often admired the literary style and quality of the *Coast Pilot* that it might be well here to quote from it. In the first place, the compilers of this book are cynical men. They know that they are writing for morons, that if by any effort their descriptions can be misinterpreted or misunderstood by the reader, that effort will be made. These writers have a contempt for almost everything. They would like an ocean and a coastline unchanging and unchangeable; lights and buoys that do not rust and wash away; winds and storms that come at specified times; and, finally, reasonably intelligent men to read their instructions. They are gratified in none of these desires. They try to write calmly and objectively, but now and then a little bitterness creeps in, particularly when they deal with Mexican lights, buoys, and port facilities. The following quotation is from H. O. No. 84, "Sailing Directions for the West Coasts of Mexico and Central America, 1937, Corrections to January 1940," page 125, under "La Paz Harbor."

> **La Paz Harbor** is that portion of La Paz Channel between the eastern end of El Mogote and the shore in the vicinity of La Paz. El Mogote is a low, sandy, bush-covered peninsula, about 6 miles long, east and west, and 1½ miles wide at its widest part, that forms the northern side of Ensenada de Anpe, a large lagoon. This lagoon lies in a low plain that is covered with a thick growth of trees, bushes, and cactus. The water is shoal over the greater part of the lagoon, but a channel in which there are depths of 2 to 4 fathoms leads from La Paz Harbor to its northwestern part.

La Paz Harbor is ½ to ¾ mile wide, but it is nearly filled with shoals through which there is a winding channel with depths of 3 to 4 fathoms. A shoal with depths of only 1 to 8 feet over it extends northward from the eastern end of El Mogote to within 400 yards of Prieta Point and thus protects La Paz Harbor from the seas caused by northwesterly winds.

La Paz Channel, leading between the shoal just mentioned and the mainland, and extending from Prieta Point to abreast the town of La Paz, has a length of about 3½ miles and a least charted depth of 3¼ fathoms, but this depth can not be depended upon. Vessels of 13-foot draft may pass through the channel at any stage of the tide. The channel is narrow, with steep banks on either side, the water in some places shoaling from 3 fathoms to 3 or 4 feet within a distance of 20 yards. The deep water of the channel and the projecting points of the shoals on either side can readily be distinguished from aloft. In 1934 the controlling depth in the channel was reported to be 16 feet.

A 9-foot channel, frequently used by coasters, leads across the shoal bank and into La Paz Channel at a position nearly 1 mile south-southeastward of Prieta Point. Caymancito Rock, on the eastern side of La Paz Channel, bearing 129°, leads through this side channel.

Beacons—Off Prieta Point, at the entrance to the channel leading to La Paz, there are three beacons consisting of lengths of 3-inch pipe driven into the bottom and extending only a few feet above the surface of the water. They are difficult to make out at high tide in the daytime, and are not lighted at night [here the hatred creeps in subtly].

Light Beacons—Three pairs of concrete range beacons, from each of which a light is shown, mark La Paz Channel. The outer range is situated on the shore near the entrance to the channel, about 1 mile southeastward of Prieta Point; the middle range is on a hillside about ¼ mile south-southeastward of Caymancito Rock; and the inner range is situated about ¾ mile northeastward of the municipal pier at La Paz. . . .

Harbor Lights—A light is shown from a wooden post 18 feet high and another from a post 20 feet high on the north and south ends, respectively, of the T-head of the municipal pier at La Paz. . . .

Anchorage—Vessels waiting for a pilot can anchor southward of Prieta Point in depths of 7 to 10 fathoms. Anchorage can also be taken northward of El Mogote, but it is exposed to wind and sea. . . .

The best berth off the town is 200 to 300 yards westward of the pier in a depth of about 3½ fathoms, sand. . . .

Pilotage is compulsory for all foreign merchant vessels. Pilots come out in a small motor launch carrying a white flag on which is the letter P, and board incoming vessels in the vicinity of Prieta Point. Although pilots will take vessels in at night, it is not advisable to attempt to enter the harbor after dark.

This is a good careful description by men whose main drive is toward accuracy, and they must be driven frantic as man and tide and wave undermine their work. The shifting sands of the channel; the three-inch pipe driven into the bottom; the T-head municipal pier with its lights on wooden posts, none of which has been there for some time; and, last, their conviction that the pilots cannot find the channel at night, make for their curious, cold, tactful statement. We trust these men. They are controlled, and only now and then do their nerves break and a cry of pain escape them thus, in the "Supplement" dated 1940:

Page 109, Line 1, for "*LIGHTS*" read "*LIGHT*" and for "*TWO LIGHTS ARE*" read "*WHEN THE CANNERY IS IN OPERATION, A LIGHT IS.*"

Or again:

Page 149, Line 2, after "*line*" add: "*two piers project inward from this mole, affording berths for vessels and, except alongside these two piers, the mole is foul with debris and wrecked cranes.*"

These coast pilots are constantly exasperated; they are not happy men. When anything happens they are blamed, and their writing takes on an austere tone because of it. No matter how hard they work, the restlessness of nature and the carelessness of man are always two jumps ahead of them.

We ran happily up under Prieta Point as suggested, and dropped anchor and put up the American flag and under it the yellow quarantine flag. We would have liked to fire a gun, but we had only the ten-gauge shotgun, and its hammer was rusted down. It was only for a show of force anyway; we had never intended it for warlike purposes. And then we sat and waited. The site was beautiful—the highland of Prieta Point and a tower on the hillside. In the distance we could see the beach of La Paz, and it really looked like a Hollywood production, the fine, low buildings close down to the water and trees flanking them and a colored bandstand on the water's edge. The little canoes of Nayarit sailed by, and the sea was ruffled with a fair breeze. We took some color motion pictures of the scene, but they didn't come out either.

After what seemed a very long time, the little launch mentioned in the *Coast Pilot* started for us. But it had no white flag with the letter "P." Like the municipal pier, that was gone. The pilot, an elderly man in a business suit and a dark hat, came stiffly aboard. He had great dignity. He refused a drink, accepted cigarettes, took his position at the wheel, and ordered us on grandly. He looked like an admiral in civilian clothes. He governed Tex with a sensitive hand—a gentle push forward against the air meant "ahead." A flattened hand patting downward signified "slow." A quick thumb over the shoulder, "reverse." He was not a talkative man, and he ran us through the channel with ease, hardly scraping us at all, and signaled our anchor down 250 yards westward of the municipal pier—if there had been one—the choicest place in the harbor.

La Paz grew in fascination as we approached. The square, iron-shuttered colonial houses stood up right in back of the beach with rows of beautiful trees in front of them. It is a lovely place. There is a broad promenade along the water lined with benches, named for dead residents of the city, where one may rest oneself.

Soon after we had anchored, the port captain, customs man, and agent came aboard. The captain read our papers, which complimented us rather highly, and was so impressed that he immediately assigned us an armed guard—or, rather, three shifts of armed guards—to protect us from theft. At first we did not like this, since we had to pay these men, but we soon found the wisdom of it. For we swarmed with visitors from morning to night; little boys clustered on us like flies, in the rigging and on the deck. And although we were infested and crawling with very poor people and children, we lost nothing; and this in spite of the fact that there were little gadgets lying about that any one of us would have stolen if we had had the chance. The guards simply kept our visitors out of the galley and out of the cabin. But we do not think they prevented theft, for in other ports where we had no guard nothing was stolen.

The guards, big pleasant men armed with heavy automatics, wore uniforms that were starched and clean, and they were helpful and sociable. They ate with us and drank coffee with us and told us many valuable things about the town. And in the end we gave each of them a carton of cigarettes, which seemed valuable to them. But they were the reverse of what is usually thought and written of Mexican soldiers—they were clean, efficient, and friendly.

With the port captain came the agent, probably the finest invention of all. He did everything for us, provisioned us, escorted us, took us to dinner, argued prices for us in local stores, warned us about some places and recommended others. His fee was so small that we doubled it out of pure gratitude.

As soon as we were cleared, Sparky and Tiny and Tex went ashore and disappeared, and we did not see them until late that night, when they came back with the usual presents: shawls and carved cow-horns and colored handkerchiefs. They were so delighted with the exchange (which was then six pesos for a dollar) that we were very soon deeply laden with curios. There were five huge stuffed sea-turtles in one bunk alone, and Japanese toys, combs from New England, Spanish shawls from New Jersey, machetes from Sheffield and New York; but all of them, from having merely lived a while in La Paz, had taken on a definite Mexican flavor. Tony, who does not trust foreigners, stayed aboard, but later even he went ashore for a while.

The tide was running out and the low shore east of the town was beginning to show through the shallow water. We gathered our paraphernalia and started for the beach, expecting and finding a fauna new to us. Here on the flats the water is warm, very warm, and there is no wave-shock. It would be strange indeed if, with few exceptions of ubiquitous animals, there should not be a definite change. The base of this flat was of rubble in which many knobs and limbs of old coral were imbedded, making an easy hiding place for burrowing animals. In rubber boots we moved over the flat uncovered by the dropping tide; a silty sand made the water obscure when a rock or a piece of coral was turned over. And as always when one is collecting, we were soon joined by a number of small boys. The very posture of search, the slow movement with the head down, seems to draw people. "What did you lose?" they ask.

"Nothing."

"Then what do you search for?" And this is an embarrassing question. We search for something that will seem like truth to us; we search for understanding; we search for that principle which keys us deeply into the pattern of all life; we search for the relations of things, one to another, as this young man searches for a warm light in his wife's eyes and that one for the hot warmth of fighting. These little boys and young men on the tide flat do not even know that they search for such things too. We say to them, "We are looking for curios, for certain small animals."

Then the little boys help us to search. They are ragged and dark and each one carries a small iron harpoon. It is the toy of La Paz, owned and treasured as tops or marbles are in America. They poke about the rocks with their little harpoons, and now and then a lazing fish which blunders too close feels the bite of the iron.

There is a small ghost shrimp which lives on these flats, an efficient little fellow who lives in a burrow. He moves very rapidly, and is armed with claws which can pinch painfully. He retires backward into his hole, so that to come at him from above is to invite his weapons. The little boys solved the problem for us. We offered ten centavos for each one they took. They dug into the rubble and old coral until they got behind the ghost shrimp in his burrow, then, prodding, they drove him outraged from his hole. Then they banged him good to reduce his pinching power. We refused to buy the banged-up ones—they had to get us lively ones. Small boys are the best collectors in the world. Soon they worked out a technique for catching the shrimps with only an occasionally pinched finger, and then the ten-centavo pieces began running out, and an increasing cloud of little boys brought us specimens. Small boys have such sharp eyes, and they are quick to notice deviation. Once they know you are generally curious, they bring amazing things. Perhaps we only practice an extension of their urge. It is easy to remember when we were small and lay on our stomachs beside a tide pool and our minds and eyes went so deeply into it that size and identity were lost, and the creeping hermit crab was our size and the tiny octopus a monster. Then the waving algae covered us and we hid under a rock at the bottom and leaped out at fish. It is very possible that we, and even those who probe space with equations, simply extend this wonder.

Among small-boy groups there is usually a stupid one who understands nothing, who brings dull things, rocks and pieces of weed, and pretends that he knows what he does. When we think of La Paz, it is always of the small boys that we think first, for we had many dealings with them on many levels.

The profile of this flat was easy to get. The ghost shrimps, called "*langusta,*" were quite common; our enemy the stinging worm was about to make us careful of our fingers; the big brittle-stars were there under the old coral, but not in such great masses as at Espíritu Santo. A number of sponges clung to the stones, and small decorated crabs skulked in the interstices. Beautiful purple polyclad worms crawled over lawns of purple tunicates; the giant oyster-like hacha[5] was not often found, but we took a few specimens. There were several growth forms of the common corals[6]; the larger and handsomer of the two slim asteroids[7]; anemones of at least three types; some club urchins and snails and many hydroids.

Some of the exposed snails were so masked with forests of algae and hydroids that they were invisible to us. We found a worm-like fixed gastropod,[8] many bivalves, including the long peanut-shaped boring clam[9]; large brilliant-orange nudibranchs; hermit crabs; mantids; flatworms which seemed to flow over the rocks like living gelatin; sipunculids; and many limpets. There were a few sun-stars, but not so many or so large as they had been at Cape San Lucas.

The little boys ran to and fro with full hands, and our buckets and tubes were soon filled. The ten-centavo pieces had long run out, and ten little boys often had to join a club whose center and interest was a silver peso, to be changed and divided later. They seemed to trust one another for the division. And certainly they felt there was no chance of their being robbed. Perhaps they are not civilized and do not know how valuable money is. The poor little savages seem not to have learned the great principle of cheating one another.

[5] *Pinna* sp.
[6] *Porites.*
[7] *Phataria.*
[8] *Aletes,* or similar.
[9] *Lithophaga plumula,* or similar.

The population of small boys at La Paz is tremendous, and we had business dealings with a good part of it. Hardly had we returned to the *Western Flyer* and begun to lay out our specimens when we were invaded. Word had spread that there were crazy people in port who gave money for things a boy could pick up on the rocks. We were more than invaded—we were deluged with small boys bearing specimens. They came out in canoes, in flatboats, some even swam out, and all of them carried specimens. Some of the things they brought we wanted and some we did not want. There were hurt feelings about this, but no bitterness. Battalions of boys swarmed back to the flats and returned again. The second day little boys came even from the hills, and they brought every conceivable living thing. If we had not sailed the second night they would have swamped the boat. Meanwhile, in our dealings on shore, more small boys were involved. They carried packages, ran errands, directed us (mostly wrongly), tried to anticipate our wishes; but one boy soon emerged. He was not like the others. His shoulders were not slender, but broad, and there was a hint about his face and expression that seemed Germanic or perhaps Anglo-Saxon. Whereas the other little boys lived for the job and the payment, this boy created jobs and looked ahead. He did errands that were not necessary, he made himself indispensable. Late at night he waited, and the first dawn saw him on our deck. Further, the other small boys seemed a little afraid of him, and gradually they faded into the background and left him in charge.

Some day this boy will be very rich and La Paz will be proud of him, for he will own the things other people must buy or rent. He has the look and the method of success. Even the first day success went to his head, and he began to cheat us a little. We did not mind, for it is a good thing to be cheated a little; it causes a geniality and can be limited fairly easily. His method was simple. He performed a task, and then, getting each of us alone, he collected for the job so that he was paid several times. We decided we would not use him any more, but the other little boys decided even better than we. He disappeared, and later we saw him in the town, his nose and lips heavily bandaged. We had the story from another little boy. Our financial wizard told the others that he was our sole servant and that we had said that they weren't to come around any more. But they discovered the lie and waylaid him and beat him very badly. He wasn't a very brave little boy, but he will be a rich one because he wants to. The others wanted only sweets or a new handkerchief, but the aggressive little boy wishes to be rich, and they will not be able to compete with him.

On the evening of our sailing we had rather a sad experience with another small boy. We had come ashore for a stroll, leaving our boat tied to a log on the beach. We walked up the curiously familiar streets and ended, oddly enough, in a bar to have a glass of beer. It was a large bar with high ceilings, and nearly deserted. As we sat sipping our beer we saw a ferocious face scowling at us. It was a very small, very black Indian boy, and the look in his eyes was one of hatred. He stared at us so long and so fiercely that we finished our glasses and got up to go. But outside he fell into step with us, saying nothing. We walked back through the softly lighted streets, and he kept pace. But near the beach he began to pant deeply. Finally we got to the beach and as we were about to untie the skiff he shouted in panic, "*Cinco centavos!*" and stepped back as from a blow. And then it seemed that we could see almost how it was. We have been the same way trying to get a job. Perhaps the father of this little boy said, "Stupid one, there are strangers in the town and they are throwing money away. Here sits your father with a sore leg and you do nothing. Other boys are becoming rich, but you, because of your sloth, are not taking advantage of this miracle. Señor Ruiz had a cigar this afternoon and a glass of beer at the *cantina* because his fine son is not like you. When have you known me, your father, to have a cigar? Never. Now go and bring back some little piece of money."

Then that little boy, hating to do it, was burdened with it nevertheless. He hated us, just as we have hated the men we have had to ask for jobs. And he was afraid, too, for we were foreigners. He put it off as long as he could, but when we were about to go he had to ask and he made it very humble. Five centavos. It did seem that we knew how hard it had been. We gave him a peso, and then he smiled broadly and he looked about for something he could do for us. The boat was tied up, and he attacked the water-soaked knot like a terrier, even working at it with his teeth. But he was too little and he could not do it. He nearly cried then. We cast off and pushed the boat away, and he waded out to guide us as far as he could. We felt both good and bad about it; we hope his father bought a cigar and an *aguardiente,* and became mellow and said to a group of men in that little boy's hearing, "Now you take Juanito. You have rarely seen such a good son. This very cigar is a gift to his father who has hurt his leg. It is a matter of pride, my friends, to have a son like Juanito." And we hope he gave Juanito, if that was his name, five centavos to buy an ice and a paper bull with a firecracker inside.

No doubt we were badly cheated in La Paz. Perhaps the boatmen cheated us and maybe we paid too much for supplies—it is very hard to know. And besides, we were so incredibly rich that we couldn't tell, and we had no instinct for knowing when we were cheated. Here we were rich, but in our own country it was not so. The very rich develop an instinct which tells them when they are cheated. We knew a rich man who owned several large office buildings. Once in reading his reports he found that two electric-light bulbs had been stolen from one of the toilets in one of his office buildings. It hurt him; he brooded for weeks about it. "Civilization is dying," he said. "Whom can you trust any more? This little theft is an indication that the whole people is morally rotten."

But we were so newly rich that we didn't know, and besides we were a little flattered. The boatmen raised their price as soon as they saw the Sea-Cow wouldn't work, but as they said, times are very hard and there is no money.

12

MARCH 22

This was Good Friday, and we scrubbed ourselves and put on our best clothes and went to church, all of us. We were a kind of parade on the way to church, feeling foreign and out of place. In the dark church it was cool, and there were a great many people, old women in their black shawls and Indians kneeling motionless on the floor. It was not a very rich church, and it was old and out of repair. But a choir of small black children made the Stations of the Cross. They sang music that sounded like old Spanish madrigals, and their voices were shrill and sharp. Sometimes they faltered a little bit on the melody, but they hit the end of each line shrieking. When they had finished, a fine-looking young priest with a thin ascetic face and the hot eyes of fervency preached from over their heads. He filled the whole church with his faith, and the people were breathlessly still. The ugly bloody Christs and the simpering Virgins and the over-dressed saints were suddenly out of it. The priest was purer and cleaner and stronger than they. Out of his own purity he seemed to plead for them. After a long time we got up and went out of the dark cool church into the blinding white sunlight.

The streets were very quiet on Good Friday, and no wind blew in the trees, the air was full of the day—a kind of hush, as though the world awaited a little breathlessly the dreadful experiment of Christ with death and Hell; the testing in a furnace of an idea. And the trees and the hills and the people seemed to wait as a man waits when his wife is having a baby, expectant and frightened and horrified and half unbelieving.

There is no certainty that the Easter of the Resurrection will really come. We were probably literarily affected by the service and the people and their feeling about it, the crippled and the pained who were in the church, the little half-hungry children, the ancient women with eyes of patient tragedy who stared up at the plaster saints with eyes of such pleading. We liked them and we felt at peace with them. And strolling slowly through the streets we thought a long time of these people in the church. We thought of the spirits of kindness which periodically cause them to be fed, a little before they are dropped back to hunger. And we thought of the good men who labored to cure them of disease and poverty.

And then we thought of what they are, and we are—products of disease and sorrow and hunger and alcoholism. And suppose some all-powerful mind and will should cure our species so that for a number of generations we would be healthy and happy? We are the products of our disease and suffering. These are factors as powerful as other genetic factors. To cure and feed would be to change the species, and the result would be another animal entirely. We wonder if we would be able to tolerate our own species without a history of syphilis and tuberculosis. We don't know.

Certain communicants of the neurological conditioning religions practiced by cowardly people who, by narrowing their emotional experience, hope to broaden their lives, lead us to think we would not like this new species. These religionists, being afraid not only of pain and sorrow but even of joy, can so protect themselves that they seem dead to us. The new animal resulting from purification of the species might be one we wouldn't like at all. For it is through struggle and sorrow that people are able to participate in one another—the heartlessness of the healthy, well-fed, and unsorrowful person has in it an infinite smugness.

On the water's edge of La Paz a new hotel was going up, and it looked very expensive. Probably the airplanes will bring week-enders from Los Angeles before long, and the beautiful poor bedraggled old town will bloom with a Floridian ugliness.

Hearing a burst of chicken voices, we looked over a mud wall and saw that there were indeed chickens in the yard behind it. We asked then of a woman if we might buy several. They could be sold, she said, but they were not what one calls "for sale." We entered her yard. One of the proofs that they were not for sale was that we had to catch them ourselves. We picked out two which looked a little less muscular than the others, and went for them. Whatever has been said, true or not, of the indolence of the Lower Californian is entirely untrue of his chickens. They were athletes, highly trained both in speed and in methods of escape. They could run, fly, and, when cornered, disappear entirely and re-materialize in another part of the yard. If the owner did not want to catch them, that hesitancy was not shared by the rest of La Paz. People and children came from everywhere; a mob collected, first to give excited advice and then to help. A pillar of dust arose out of that yard. Small boys hurled themselves at the chickens like football-players. We were bound to catch them sooner or later, for as one group became exhausted, another took up the chase. If we had played fair and given those chickens rest periods, we would never have caught them. But by keeping at them, we finally wore them down and they were caught, completely exhausted and almost shorn of their feathers. Everyone in the mob felt good and happy then and we paid for the chickens and left.

On board it was Sparky's job to kill them, and he hated it. But finally he cut their heads off and was sick. He hung them over the side to bleed and a boat came along and mashed them flat against our side. But even then they were tough. They had the most highly developed muscles we have ever seen. Their legs were like those of ballet dancers and there was no softness in their breasts. We stewed them for many hours and it did no good whatever. We were sorry to kill them, for they were gallant, fast chickens. In our country they could easily have got scholarships in one of our great universities and had collegiate careers, for they had spirit and fight and, for all we know, loyalty.

On the afternoon tide we were to collect on El Mogote, a low sandy peninsula with a great expanse of shallows which would be exposed at low tide. The high-tide level was defined by a heavy growth of mangrove. The area was easily visible from our anchorage, and the sand was smooth and not filled with rubble or stones or coral. A tall handsome boy of about nineteen had been idling about the *Western Flyer*. He had his own canoe, and he offered to paddle us to the tide flats. This boy's name was Raúl Velez; he spoke some English and was of great service to us, for his understanding was quick and he helped valuably at the collecting. He told us the local names of many of the animals we had taken; "cornuda" was the hammer-head shark; "barco," the red snapper; "caracol," and also "burral," all snails in general, but particularly the large conch. Urchins were called "erizo" and sea-fans, "abanico." "Bromas" were barnacles and "hacha" the pinna, or large clam.

The sand flats were very interesting. We dug up a number of

Dentaliums of two species, the first we had found. These animals, which look like slender curved teeth, belong to a small class of mollusks, little known popularly, called "tooth shells."

On the shallow bottom, attached to very small stones, we found little anemones of three types. There were also sand anemones,[1] in long filthy-looking gray cases when they were dug out. But when they were imbedded in the bottom and expanded, they looked like lovely red and purple flowers. A great number of small black cucumbers of a type we had not taken crawled on the bottom, as well as one large pepper-and-salt cucumber. We found many heart-urchins, two species of the ordinary ophiurans (brittle-stars), and one burrowing ophiuran. Sponges and tunicates were fastened to the insecure footing of very small stones, but since there is probably very little churning of water on the tide flats, they were safe enough. There were flat worms of several species; stinging worms, peanut worms, echiuroid worms, and what in the collecting notes are listed rather tiredly as "worms." We took one specimen of the sea-whip, a rather spectacular colony of animals looking exactly like a long white whip. The lower portion is a horny stalk and the upper part consists of zooids carrying on their own life processes but connected by a series of canals which unite their body cavities with the main stalk.

As the tide came up we moved upward in the intertidal toward the mangrove trees, and the foul smell of them reached us. They were in bloom, and the sharp sweet smell of their flowers, combined with the filthy odor of the mud about their roots, was sickening. But they are fascinating to look into. Huge hermit crabs seem to live among their stilted roots; the black mud, product of the root masses, swarms as a meeting place for land and sea animals. Flies and insects in great numbers crawl and buzz about the mud, and the scavenging hermit crabs steal secretly in and out and even climb into the high roots.

We suppose it is the combination of foul odor and the impenetrable quality of the mangrove roots which gives one a feeling of dislike for these salt-water-eating bushes. We sat quietly and watched the moving life in the forests of the roots, and it seemed to us that there was stealthy murder everywhere. On the surf-swept

[1] *Cerianthus.*

rocks it was a fierce and hungry and joyous killing, committed with energy and ferocity. But here it was like stalking, quiet murder. The roots gave off clicking sounds, and the odor was disgusting. We felt that we were watching something horrible. No one likes the mangroves. Raúl said that in La Paz no one loved them at all.

On the level flats the tide covers the area very quickly. We waded out to a wrecked boat lying turned over on the sand, and took a number of barnacles from the rotten wood and even from the rusted engine. It was a good rich collecting day, and it had been a curiously emotional day beginning with the church. Sometimes one has a feeling of fullness, of warm wholeness, wherein every sight and object and odor and experience seems to key into a gigantic whole. That day even the mangrove was part of it. Perhaps among primitive peoples the human sacrifice has the same effect of creating a wholeness of sense and emotion—the good and bad, beautiful, ugly, and cruel all welded into one thing. Perhaps a whole man needs this balance. And we had been as excited at finding the Dentaliums as though they had been nuggets of gold.

Raúl had a La Paz harpoon in his canoe, and we bought it from him, hoping to bring it home. It was a shaft of iron with a ring on one end for the line and a point and hinged barb at the other. A little circle of cord holds the barb against the shaft until the friction of the flesh of the victim pushes the cord free and allows the barb to open out inside the flesh. We wanted to keep this harpoon, but we lost it in a manta ray later. At this reading, there are many manta rays in the Gulf cruising about with our harpoons in their hides.

We also wanted one of the Nayarit canoes to take back, for they are light and of shallow draft, ideal for collecting in the lagoons and seaworthy even in rough water. But no one would sell a canoe. They came from too far away and were too well loved. Some very old ones were solid with braces and patches.

It was dusk when we came back to the *Western Flyer*, and the deck was filled with waiting little boys holding mashed and mangled specimens of all kinds. We bought what we needed and then we bought a lot of things we didn't need. The boys had waited a long time for us, under the stern eye of our military man. And it was interesting to see how our soldier loved the ragged little boys of La Paz. When they got out of hand or ran too fast over the deck, he cautioned them, but there was none of the bluster of the policeman. And had we not been just to the little boys, he would have joined them; for they were his people and our great wealth would not have deflected him from them. He wore his automatic, but it was only a badge with no show of force about it, and when he entered the galley or sat down with us he removed the gun belt and hung it up. We liked the tone of voice he used on the boys. It had dignity and authority but no bullying quality, and the boys of the town seemed to respect him without fearing him.

Once when a little boy practiced the most ancient trick in the list of boy skulduggery—that of removing a specimen and selling it again—the soldier spoke to him shortly with contempt, and that boy lost his standing and even his friends.

One boy had, on a light harpoon, a fish which looked something like the puffers—a gray and black fish with a large flat head. When we wished to buy it he refused, saying that a man had commissioned him to get this fish and he was to receive ten centavos for it because the man wanted to poison a cat. This was the *botete,* and our first experience with it. It is thought in La Paz that the poison concentrates in the liver and this part is used for poisoning small animals and even flies. We did not make this test, but we found *botete* everywhere in the warm shallow waters of the Gulf. Probably he is the most prevalent fish of all in lagoons and eel-grass flats. He lies on the bottom, and his marking makes him nearly invisible. Sometimes he lies in a small cleared place in eel-grass or in a slight depression on the silt bottom which indicates, but does not prove, that he has a fairly permanent resting-place to which he returns. When one is wading in the shallows, *botete* lies quiet until he is nearly stepped on before he streaks away, drawing a cloud of disturbed mud after him.

In the press of collecting and preserving, we neglected to dissect the stomach of this fish, so we do not know what he eats.

The literature on *botete* is scattered and hard to come by. Members of his genus, having his poisonous qualities, are distributed all over the world where there are shallows of warm water. Since this fish is very dangerous to eat and is so widely found, it is curious that so little has been written about it. Eating him almost invariably causes death in agony. If he were rare, it would be understandable why he has been so little discussed. But more has been written about some of the seldom-seen fishes of the great depths than of this deadly little *botete*. We were fascinated with him and took a number of specimens. Following are some of the few reports available on his nature and misdemeanors. We still do not know whether he kills flies.

From Herre[2] we learn that "In at least two or three of the suborders the flesh nearly always is not only thin, hard, often bitter and usually unpalatable, but also contains poisonous alkaloids. These produce the disease known as ciguatera, in which the nervous system is attacked and violent gastric disturbances, paralysis, and death may follow."

On page 423 he discusses the Balistidae, or trigger-fish such as the Gulf puerco: "Although seen in fish markets throughout the Orient, none of the Balistidae are much used as human food. In some localities of the Philippines, those of moderate size are eaten, but their sale here should be forbidden as their flesh is always more or less poisonous. In such places as Cuba and Mauritius they are not allowed in the markets as they are known to cause ciguatera.

"Francis Day says (*Fishes of India,* 1878, p. 686): 'Dr. Meunier, at Mauritius, considers that the poisonous flesh acts primarily on the nervous tissue of the stomach, causing violent spasms of that organ and, shortly afterwards, of all the muscles of the body. The frame becomes wracked with spasms, the tongue thickened, the eye fixed, the breathing laborious, and the patient expires in a paroxysm of extreme suffering. The first remedy to be given is a strong emetic, and subsequently oils and demulcents to allay irritability.'

"In his account of the backboned animals of Abyssinia Rüppel states that *Balistes flavomarginatus* is very common in the Red Sea at Djetta, where it is often brought to market, although only pilgrims who do not know the quality of the flesh will buy it. He goes on to say that as a whole the Balistidae not only have a bad taste, but also are unwholesome as food."

Referring to the Tetraodontidae, page 479, Herre uses the name *batete,* or *botete,* as used in most Philippine languages. "This dangerous group of fishes," he says, "is widely distributed in warm seas all over the world and is common throughout the Philippines. Although most people are more or less aware of the poisonous

[2] "Poisonous and Worthless Fishes: An Account of the Philippine Plectognaths," *Phil. Journ. Sci.*, Vol. 25 (4), p. 415.

properties of the flesh, it is eaten in practically every Philippine fishing village and not a year goes by without several deaths from this cause.

"A Japanese investigator (I have been unable to obtain a copy of his paper, which appeared in *Archiv für Pathologie und Pharmacologie*) has studied carefully the alkaloid present in the flesh of the Tetraodontidae and finds it to be very near to muscarine, the active poisonous principle of *Amanita muscaria* and other fungi. It is a tasteless, odorless, and very poisonous crystalline alkaloid."

He goes on to state that the natives consider the gall-bladder, the milt, and the eggs to be particularly poisonous. But in La Paz it was the liver which was thought to be the most poisonous part. Only the liver was used to poison animals and flies, although this might be because the liver was more attractive as bait than other portions.

Herre continues on page 488 concerning *Tetraodon*: "The *Medical Journal of Australia* under the date of December 1, 1923, tells of two Malays who ate of a species of Tetraodon although warned of the danger. They ate at noon with no serious effects, but on eating some for supper they were taken violently ill, one dying in an hour, the other about three hours later." Of the Diodontidae, page 503 (the group to which the puffer fish belong): "The fishes of this family have a well-deserved reputation for being poisonous and their flesh should never be eaten."

Botete is sluggish, fairly slow, unarmored, and not very clever at either concealment, escape, or attack. It is amusing but valueless to speculate anthropomorphically in the chicken-egg manner regarding the relationship between his habits and his poison. Did he develop poison in his flesh as a protection in lieu of speed and cleverness, or being poisonous and quite unattractive, was he able to "let himself go," to abandon speed and cleverness? The protected human soon loses his power of defense and attack. Perhaps *botete*, needing neither brains nor tricks nor techniques to protect himself except from a man who wants to poison a cat, has become a frump.

In the evening Tiny returned to the *Western Flyer*, having collected some specimens of *Phthirius pubis*, but since he made no notes in the field, he was unable or unwilling to designate the exact collecting station. His items seemed to have no unusual qualities but to be members of the common species so widely distributed throughout the world.

We were to sail in the early morning, and that night we walked a little in the dim-lighted streets of La Paz. And we wondered why so much of the Gulf was familiar to us, why this town had a "home" feeling. We had never seen a town which even looked like La Paz, and yet coming to it was like returning rather than visiting. Some quality there is in the whole Gulf that trips a trigger of recognition so that in fantastic and exotic scenery one finds oneself nodding and saying inwardly, "Yes, I know." And on the shore the wild doves mourn in the evening and then there comes a pang, some kind of emotional jar, and a longing. And if one followed his whispering impulse he would walk away slowly into the thorny brush following the call of the doves. Trying to remember the Gulf is like trying to re-create a dream. This is by no means a sentimental thing, it has little to do with beauty or even conscious liking. But the Gulf does draw one, and we have talked to rich men who own boats, who can go where they will. Regularly they find themselves sucked into the Gulf. And since we have returned, there is always in the backs of our minds the positive drive to go back again. If it were lush and rich, one could understand the pull, but it is fierce and hostile and sullen. The stone mountains pile up to the sky and there is little fresh water. But we know we must go back if we live, and we don't know why.

Late at night we sat on the deck. They were pumping water out of the hold of the trading boat, preparing her to float and flounder away to Guaymas for more merchandise. But La Paz was asleep; not a soul moved in the streets. The tide turned and swung us around, and in the channel the ebbing tide whispered against our hull and we heard the dogs of La Paz barking in the night.

13

MARCH 23

We sailed in the morning. The mustached old pilot came aboard and steered us out, then bowed deeply and stepped into the launch which had followed us. The sea was calm and very blue, almost black-blue, as we turned northward along the coast. We wished to stop near San José Island as our next collecting station. It was good to be under way again and good to be out from under the steady eyes of those ubiquitous little boys who waited interminably for us to do something amusing.

In mid-afternoon we came to anchorage at Amortajada Bay on the southwest tip of San José Island. A small dark islet had caught our attention as we came in. For although the day was bright this islet, called Cayo on the map, looked black and mysterious. We had a feeling that something strange and dark had happened there or that it was the ruined work of men's hands. Cayo is only a quarter of a mile long and a hundred yards wide. Its northern end is a spur and its southern end a flat plateau about forty feet high. Even in the distance it had a quality which we call "burned." One knows there will be few animals on a "burned" coast; that animals will not like it, will not be successful there. Even the algae will be like lost colonists. Whether or not this is the result of a deadly chemistry we cannot say. But we can say that it is possible, after long collecting, to recognize a shore which is "burned" even if it is so far away that details cannot be seen.

Cayo lay about a mile and a half from our anchorage and seemed to blacken even the air around it. This was the first time that the Sea-Cow could have been of great service to us. It was for just such occasions that we had bought it. We were kind to it that day—selfishly of course. We said nice things about it and put it tenderly on the stern of the skiff, pretending to ourselves that we expected it to run, that we didn't dream it would not run. But it would not. We rowed the boat—and the Sea-Cow—to Cayo Islet. There is so much that is strange about this islet that we will set much of it down. It is nearly all questions, but perhaps someone reading this may know the answers and tell us. There is no landing place; all approaches are strewn with large sea-rounded boulders which even in fairly still water would beat the bottom out of a boat. On its easterly side, the one we approached, a cliff rises in back of a rocky beach and there are a number of shallow caves in the cliffside. Set in the great boulders in the intertidal zone there are large iron rings and lengths of big chain, but so rusted and disintegrated that they came off in our hands. Also, set in the cliff six to eight feet above the beach, are other iron rings with loops eight inches in diameter. They look very old, but the damp air of the Gulf and the rapid oxidation caused by it make it impossible to say exactly how old they are. In the shallow caves in the cliff there were evidences of many fires' having been built, and piled about the fireplaces, some old and some fresh, were not only thousands of clam-shells but turtle-shells also, as though these animals had been brought here to be smoked. A heap of fairly fresh diced turtle-meat lay beside one of the fireplaces. The mysterious quality of all this lies here. There are no clams in this immediate vicinity and turtles do not greatly abound. There is no wood whatever on the island with which to build fires; it would have to be brought here. There is no water whatever. And once arrived, there is no anchorage. Why people would bring clams and turtles and wood and water to an islet where there was no protection we do not know. A mile and a half away they could have beached easily and have found both wood and water. It is a riddle we cannot answer, just as we can think of no reason for the big iron rings. They could not have been for fastening a big boat to, since there is no safe water for a boat to lie in and no cove protection from wind and storm. We are very curious about this. We climbed the cliff by a trail that was well beaten in a crevice and on the flat top found a sparse growth of brown grass and some cactus. Nothing more. On the southernmost end of the cliff sat one large black crow who shrieked at us with dislike, and when we approached flew off and disappeared in the direction of San José Island.

The cliffs were light buff in color, and the grass light brown. It is impossible to say why distance made Cayo look black. Boulders and fixed stones of the reef were of a reddish igneous rock and the island, like the whole region, was volcanic in origin.

Collecting on the rocks we found, as we knew we would, a sparse and unhappy fauna. The animals were very small. *Heliaster,* the sun-star of which there were a few, was small and pale in color. There were anemones, a few sea-cucumbers, and a few sea-rabbits. The one animal which seemed to like Cayo was Sally Lightfoot. These beautiful crabs crawled on the rocks and dominated the life of the region. We took a few *Aletes* (worm-like snails) and some serpulid worms, two or three types of snails, and a few isopods and beach-hoppers.

The tide came up and endangered our boat, which we had balanced on top of a boulder, and we rowed back toward the *Western Flyer,* one of us in the stern pulling with a quiet fury on the starting rope of the Sea-Cow. We wished we had left it dangling by its propeller on one of the cliff rings, and its evil and mysterious magneto would have liked that too.

As soon as we pulled away, Cayo looked black again, and we hope someone can tell us something about this island.

Back on the *Western Flyer* we asked Tex to take the Sea-Cow apart down to the tiniest screw and to find out in truth, once for all, whether its failure were metaphysical or something which could be fixed. This he did, under a deck-light. When he put it together and attached it to the boat, it ran perfectly and he went for a cruise with it. Now at last we felt we had an outboard motor we could depend on.

We were anchored quite near San José Island and that night we were visited by little black beetle-like flies which bit and left a stinging, itching burn. Covering ourselves did not help, for they crawled down inside our bedding and bit us unmercifully. Being unable to sleep, we talked and Tiny told us a little of his career, which, if even part of it is true, is one of the most decoratively disreputable sagas we have ever heard. It is with sadness that we do not include some of it, but certain members of the general public are able to keep from all a treatise on biology unsurpassed in our experience. The great literature of this kind is kept vocal by the combined efforts of Puritans and postal regulations, and so the saga of Tiny must remain unwritten.

14

MARCH 24, EASTER SUNDAY

The beach was hot and yellow. We swam, and then walked along on the sand and went inland along the ridge between the beach and a large mangrove-edged lagoon beyond. On the lagoon side of the ridge there were thousands of burrows, presumably of large land-crabs, but it was hopeless to dig them out. The shores of the lagoon teemed with the little clicking bubbling fiddler crabs and estuarian snails. Here we could smell the mangrove flowers without the foul root smell, and the odor was fresh and sweet, like that of new-cut grass. From where we waded there was a fine picture, still reflecting water and the fringing green mangroves against the burnt red-brown of the distant mountains, all like some fantastic Doré drawing of a pressed and embattled heaven. The air was hot and still and the lagoon rippleless. Now and then the surface was ringed as some lagoon fish came to the air. It was a curious quiet resting-place and perhaps because of the quiet we heard in our heads the children singing in the church at La Paz. We did not collect strongly or very efficiently, but rather we half dozed through the day, thinking of old things, each one in himself. And later we discussed manners of thinking and methods of thinking, speculation which is not stylish any more. On a day like this the mind goes outward and touches in all directions. We discussed intellectual methods and approaches, and we thought that through inspection of thinking technique a kind of purity of approach might be consciously achieved—that non-teleological or "is" thinking might be substituted in part for the usual cause-effect methods.

The hazy Gulf, with its changes of light and shape, was rather like us, trying to apply our thoughts, but finding them always pushed and swayed by our bodies and our needs and our satieties. It might be well here to set down some of the discussions of non-teleological thinking.

During the depression there were, and still are, not only destitute but thriftless and uncareful families, and we have often heard it said that the country had to support them because they were shiftless and negligent. If they would only perk up and be somebody everything would be all right. Even Henry Ford in the depth of the depression gave as his solution to that problem, "Everybody ought to roll up his sleeves and get to work."

This view may be correct as far as it goes, but we wonder what would happen to those with whom the shiftless would exchange places in the large pattern—those whose jobs would be usurped, since at that time there was work for only about seventy percent of the total employable population, leaving the remainder as government wards.

This attitude has no bearing on what might be or could be if so-and-so happened. It merely considers conditions "as is." No matter what the ability or aggressiveness of the separate units of society, at that time there were, and still there are, great numbers necessarily out of work, and the fact that those numbers comprised the incompetent or maladjusted or unlucky units is in one sense beside the point. No causality is involved in that; collectively it's just "so"; collectively it's related to the fact that animals produce more offspring than the world can support. The units may be blamed as individuals, but as members of society they cannot be blamed. Any given individual very possibly may transfer from the underprivileged into the more fortunate group by better luck or by improved aggressiveness or competence, but all cannot be so benefited whatever their strivings, and the large population will be unaffected. The seventy-thirty ratio will remain, with merely a reassortment of the units. And no blame, at least no social fault, imputes to these people; they are where they are "because" natural conditions are what they are. And so far as we selfishly are concerned we can rejoice that they, rather than we, represent the low extreme, since there must be one.

So if one is very aggressive he will be able to obtain work even under the most sub-normal economic conditions, but only because there are others, less aggressive than he, who serve in his stead as potential government wards. In the same way, the sight of a half-wit should never depress us, since his extreme, and the extreme of his kind, so affects the mean standard that we, hatless, coatless, often bewhiskered, thereby will be regarded only as a little odd.

And similarly, we cannot justly approve the success manuals that tell our high school graduates how to get a job—there being jobs for only half of them!

This type of thinking unfortunately annoys many people. It may especially arouse the anger of women, who regard it as cold, even brutal, although actually it would seem to be more tender and understanding, certainly more real and less illusionary and even less blaming, than the more conventional methods of consideration. And the value of it as a tool in increased understanding cannot be denied.

As a more extreme example, consider the sea-hare *Tethys,* a shell-less, flabby sea-slug, actually a marine snail, which may be seen crawling about in tidal estuaries, somewhat resembling a rabbit crouched over. A California biologist estimated the number of eggs produced by a single animal during a single breeding season to be more than 478 million. And the adults sometimes occur by the hundred! Obviously all these eggs cannot mature, all this potential cannot, *must not,* become reality, else the ocean would soon be occupied exclusively by sea-hares. There would be no kindness in that, even for the sea-hares themselves, for in a few generations they would overflow the earth; there would be nothing for the rest of us to eat, and nothing for them unless they turned cannibal. On the average, probably no more than the biblical one or two attain full maturity. Somewhere along the way all the rest will have been eaten by predators whose life cycle is postulated upon the presence of abundant larvae of sea-hares and other forms as food—as all life itself is based on such a postulate. Now picture the combination mother-father sea-hare (the animals are hermaphroditic, with the usual cross-fertilization) parentally blessing its offspring with these words: "Work hard and be aggressive, so you can grow into a nice husky *Tethys* like your ten-pound parent." Imagine it, the hypocrite, the illusionist, the Pollyanna, the genial liar, saying that to its millions of eggs *en masse,* with the dice loaded at such a ratio! Inevitably, 99.999 percent are destined to fall by the wayside. No prophet could foresee which specific individuals are to survive, but the most casual student could state confidently that no more than a few are likely to do so; any given individual has *almost* no chance at all—but still there is the "almost," since the race persists. And there is even a semblance of truth in the parent sea-hare's admonition, since even here, with this almost infinitesimal differential, the race is still to the swift and/or to the lucky.

What we personally conceive by the term "teleological thinking," as exemplified by the notion about the shiftless unemployed, is most frequently associated with the evaluating of causes and effects, the purposiveness of events. This kind of thinking considers changes and cures—what "should be" in the terms of an end pattern (which is often a subjective or an anthropomorphic projection); it presumes the bettering of conditions, often, unfortunately, without achieving more than a most superficial understanding of those conditions. In their sometimes intolerant refusal to face facts as they are, teleological notions may substitute a fierce but ineffectual attempt to change conditions which are assumed to be undesirable, in place of the understanding-acceptance which would pave the way for a more sensible attempt at any change which might still be indicated.

Non-teleological ideas derive through "is" thinking, associated with natural selection as Darwin seems to have understood it. They imply depth, fundamentalism, and clarity—seeing beyond traditional or personal projections. They consider events as outgrowths and expressions rather than as results; conscious acceptance as a desideratum, and certainly as an all-important prerequisite. Non-teleological thinking concerns itself primarily not with what should be, or could be, or might be, but rather with what actually "is"—attempting at most to answer the already sufficiently difficult questions *what* or *how,* instead of *why.*

An interesting parallel to these two types of thinking is afforded by the microcosm with its freedom or indeterminacy, as contrasted with the morphologically inviolable pattern of the macrocosm. Statistically, the electron is free to go where it will. But the destiny pattern of any aggregate, comprising uncountable billions of these same units, is fixed and certain, however much that inevitability may be slowed down. The eventual disintegration of a stick of wood or a piece of iron through the departure of the presumably immortal electrons is assured, even though it may be delayed by such protection against the operation of the second law of thermodynamics as is afforded by painting and rustproofing.

Examples sometimes clarify an issue better than explanations or definitions. Here are three situations considered by the two methods.

A. Why are some men taller than others?
Teleological "answer": because of the underfunctioning of the growth-regulating ductless glands. This seems simple enough. But the simplicity is merely a function of inadequacy and incompleteness. The finality is only apparent. A child, being wise and direct, would ask immediately if given this answer: "Well, why do the glands underfunction?" hinting instantly towards non-teleological methods, or indicating the rapidity with which teleological thinking gets over into the stalemate of first causes.

In the non-teleological sense there can be no "answer." There can be only pictures which become larger and more significant as one's horizon increases. In this given situation, the steps might be something like this:

(1) Variation is a universal and truly primitive trait. It occurs in any group of entities—razor blades, measuring-rods, rocks, trees, horses, matches, or men.

(2) In this case, the apropos variations will be towards shortness or tallness from a mean standard—the height of adult men as determined by the statistics of measurements, or by common-sense observation.

(3) In men varying towards tallness there seems to be a constant relation with an underfunctioning of the growth-regulating ductless glands, of the sort that one can be regarded as an index of the other.

(4) There are other known relations consistent with tallness, such as compensatory adjustments along the whole chain of endocrine organs. There may even be other factors, separately not important or not yet discovered, which in the aggregate may be significant, or the integration of which may be found to wash over some critical threshold.

(5) The men in question are taller "because" they fall in a group within which there are the above-mentioned relations. In other words, "they're tall because they're tall."

This is the statistical, or "is," picture to date, more complex than the teleological "answer"—which is really no answer at all—but complex only in the sense that reality is complex; actually simple, inasmuch as the simplicity of the word "is" can be comprehended.

Understandings of this sort can be reduced to this deep and significant summary: "It's so because it's so." But exactly the same words can also express the hasty or superficial attitude. There seems to be no explicit method for differentiating the deep and participating understanding, the "all-truth" which admits infinite change or expansion as added relations become apparent, from the shallow dismissal and implied lack of further interest which may be couched in the very same words.

B. Why are some matches larger than others?
Examine similarly a group of matches. At first they seem all to be of the same size. But to turn up differences, one needs only to measure them carefully with calipers or to weigh them with an analytical balance. Suppose the extreme comprises only a .001 percent departure from the mean (it will be actually much more); even so slight a differential we know can be highly significant, as with the sea-hares. The differences will group into plus-minus variations from a hypothetical mean to which not one single example will be found exactly to conform. Now the ridiculousness of the question becomes apparent. There is no *particular* reason. It's just so. There may be in the situation some factor or factors more important than the others: owing to the universality of variation (even in those very factors which "cause" variation), there surely *will* be, some even predominantly so. But the question as put is seen to be beside the point. The good answer is: "It's just in the nature of the beast." And this needn't imply belittlement; to have understood the "nature" of a thing is in itself a considerable achievement.

But if the size variations should be quite obvious—and especially if uniformity were to be a desideratum—then there might be a particularly dominant "causative" factor which could be searched out. Or if a person must have a stated "cause"—and many people must, in order to get an emotional understanding, a sense of relation to the situation and to give a name to the thing in order to "settle" it so it may not bother them any more—he can examine the automatic machinery which fabricates the products, and discover in it the variability which results in variation in the matches. But in doing so, he will become involved with a larger principle or pattern, the universality of variation, which has little to do with causality as we think of it.

C. Leadership.
The teleological notion would be that those in the forefront are leaders in a given movement and actually direct and consciously lead the masses in the sense that an army corporal orders "Forward

march" and the squad marches ahead. One speaks in such a way of church leaders, of political leaders, and of leaders in scientific thought, and of course there is some limited justification for such a notion.

Non-teleological notion: that the people we call leaders are simply those who, at the given moment, are moving in the direction behind which will be found the greatest weight, and which represents a future mass movement.

For a more vivid picture of this state of affairs, consider the movements of an ameba under the microscope. Finger-like processes, the pseudopodia, extend at various places beyond the confines of the chief mass. Locomotion takes place by means of the animal's flowing into one or into several adjacent pseudopodia. Suppose that the molecules which "happened" to be situated in the forefront of the pseudopodium through which the animal is progressing, or into which it will have flowed subsequently, should be endowed with consciousness and should say to themselves and to their fellows: "We are directly leading this great procession, our leadership 'causes' all the rest of the population to move this way, the mass follows the path we blaze." This would be equivalent to the attitude with which we commonly regard leadership.

As a matter of fact there are three distinct types of thinking, two of them teleological. Physical teleology, the type we have been considering, is by far the commonest today. Spiritual teleology is rare. Formerly predominant, it now occurs metaphysically and in most religions, especially as they are popularly understood (but not, we suspect, as they were originally enunciated or as they may still be known to the truly adept). Occasionally the three types may be contrasted in a single problem. Here are a couple of examples:

(1) Van Gogh's feverish hurrying in the Arles epoch, culminating in epilepsy and suicide.

Teleological "answer": Improper care of his health during times of tremendous activity and exposure to the sun and weather brought on his epilepsy out of which discouragement and suicide resulted.

Spiritual teleology: He hurried because he innately foresaw his imminent death, and wanted first to express as much of his essentiality as possible.

Non-teleological picture: Both the above, along with a good

many other symptoms and expressions (some of which could probably be inferred from his letters), were parts of his essentiality, possibly glimpsable as his "lust for life."

(2) The thyroid-neurosis syndrome.

Teleological "answer": Over-activity of the thyroid gland irritates and over-stimulates the patient to the point of nervous breakdown.

Spiritual teleology: The neurosis is causative. Something psychically wrong drives the patient on to excess mental irritation which harries and upsets the glandular balance, especially the thyroid, through shock-resonance in the autonomic system, in the sense that a purely psychic shock may spoil one's appetite, or may even result in violent illness. In this connection, note the army's acceptance of extreme homesickness as a reason for disability discharge.

Non-teleological picture: Both are discrete segments of a vicious circle, which may also include other factors as additional more or less discrete segments, symbols or maybe parts of an underlying but non-teleological pattern which comprises them and many others, the ramifications of which are *n,* and which has to do with causality only reflectedly.

Teleological thinking may even be highly fallacious, especially where it approaches the very superficial but quite common *post hoc, ergo propter hoc* pattern. Consider the situation with reference to dynamiting in a quarry. Before a charge is set off, the foreman toots warningly on a characteristic whistle. People living in the neighborhood come to associate the one with the other, since the whistle is almost invariably followed within a few seconds by the shock and sound of an explosion for which one automatically prepares oneself. Having experienced this many times without closer contact, a very naïve and unthinking person might justly conclude not only that there was a cause-effect relation, but that the whistle actually caused the explosion. A slightly wiser person would insist that the explosion caused the whistle, but would be hard put to explain the transposed time element. The normal adult would recognize that the whistle no more caused the explosion than the explosion caused the whistle, but that both were parts of a larger pattern out of which a "why" could be postulated for both, but more immediately and particularly for the whistle. Determined to chase the thing down in a cause-effect sense, an observer would have to be very wise indeed who could follow the intricacies of cause through more fundamental cause to primary cause, even in this largely man-made series about which we presumably know most of the motives, causes, and ramifications. He would eventually find himself in a welter of thoughts on production, and ownership of the means of production, and economic whys and wherefores about which there is little agreement.

The example quoted is obvious and simple. Most things are far more subtle than that, and have many of their relations and most of their origins far back in things more difficult of access than the tooting of a whistle calculated to warn bystanders away from an explosion. We know little enough even of a manmade series like this—how much less of purely natural phenomena about which also there is apt to be teleological pontificating!

Usually it seems to be true that when even the most definitely apparent cause-effect situations are examined in the light of wider knowledge, the cause-effect aspect comes to be seen as less rather than more significant, and the statistical or relational aspects acquire larger importance. It seems safe to assume that non-teleological is more "ultimate" than teleological reasoning. Hence the latter would be expected to prove to be limited and constricting except when used provisionally. But while it is true that the former is more open, for that very reason its employment necessitates greater discipline and care in order to allow for the dangers of looseness and inadequate control.

Frequently, however, a truly definitive answer seems to arise through teleological methods. Part of this is due to wish-fulfillment delusion. When a person asks "Why?" in a given situation, he usually deeply expects, and in any case receives, only a relational answer in place of the definitive "because" which he thinks he wants. But he customarily accepts the actually relational answer (it couldn't be anything else unless it comprised the whole, which is unknowable except by "living into") as a definitive "because." Wishful thinking probably fosters that error, since everyone continually searches for absolutisms (hence the value placed on diamonds, the most permanent physical things in the world) and imagines continually that he finds them. More justly, the relational picture should be regarded only as a glimpse—a challenge to consider also the rest of the relations as they are available—to envision the whole picture as well as can be done with given abilities and data. But one accepts it instead of a real "because," considers it settled, and, having named it, loses interest and goes on to something else.

Chiefly, however, we seem to arrive occasionally at definitive answers through the workings of another primitive principle: the universality of quanta. No one thing ever merges gradually into anything else; the steps are discontinuous, but often so very minute as to seem truly continuous. If the investigation is carried deep enough, the factor in question, instead of being graphable as a continuous process, will be seen to function by discrete quanta with gaps or synapses between, as do quanta of energy, undulations of light. The apparently definitive answer occurs when causes and effects both arise on the same large plateau which is bounded a great way off by the steep rise which announces the next plateau. If the investigation is extended sufficiently, that distant rise will, however, inevitably be encountered; the answer which formerly seemed definitive now will be seen to be at least slightly inadequate and the picture will have to be enlarged so as to include the plateau next further out. Everything impinges on everything else, often into radically different systems, although in such cases faintly. We doubt very much if there are any truly "closed systems." Those so called represent kingdoms of a great continuity bounded by the sudden discontinuity of great synapses which eventually must be bridged in any unified-field hypothesis. For instance, the ocean, with reference to waves of water, might be considered as a closed system. But anyone who has lived in Pacific Grove or Carmel during the winter storms will have felt the house tremble at the impact of waves half a mile or more away impinging on a totally different "closed" system.

But the greatest fallacy in, or rather the greatest objection to, teleological thinking is in connection with the emotional content, the belief. People get to believing and even to professing the apparent answers thus arrived at, suffering mental constrictions by emotionally closing their minds to any of the further and possibly opposite "answers" which might otherwise be unearthed by honest effort—answers which, if faced realistically, would give rise to a struggle and to a possible rebirth which might place the whole problem in a new and more significant light. Grant for a moment that among students of endocrinology a school of thought might arise, centering upon some belief as to etiology—upon the belief, for instance, that all abnormal growth is caused by glandular imbalance. Such a clique, becoming formalized and powerful, would tend, by scorn and opposition, to wither any contrary view which, if untrammeled, might discover a clue to some opposing "causative" factor of equal medical importance. That situation is most unlikely to arise in a field so lusty as endocrinology, with its relational insistence, but the principle illustrated by a poor example is thought nevertheless to be sound.

Significant in this connection is the fact that conflicts may arise between any two or more of the "answers" brought forth by either of the teleologies, or between the two teleologies themselves. But there can be no conflict between any of these and the non-teleological picture. For instance, in the condition called hyperthyroidism, the treatments advised by believers in the psychic or neurosis etiology very possibly may conflict with those arising out of a belief in the purely physical cause. Or even within the physical teleology group there may be conflicts between those who believe the condition due to a strictly thyroid upset and those who consider causation derived through a general imbalance of the ductless glands. But there can be no conflict between any or all of these factors and the non-teleological picture, because the latter includes them—evaluates them relationally or at least attempts to do so, or maybe only accepts them as time-place truths. Teleological "answers" necessarily must be included in the non-teleological method—since they are part of the picture even if only restrictedly true—and as soon as their qualities of relatedness are recognized. Even erroneous beliefs are real things, and have to be considered proportional to their spread or intensity. "All-truth" must embrace all extant apropos errors also, and know them as such by relation to the whole, and allow for their effects.

The criterion of validity in the handling of data seems to be this: that the summary shall say in substance, significantly and understandingly, "It's so because it's so." Unfortunately the very same words might equally derive through a most superficial glance, as any child could learn to repeat from memory the most abstruse of Dirac's equations. But to know a thing emergently and significantly is something else again, even though the understanding may be expressed in the self-same words that were used superficially. In the following example[1] note the deep significance of the emergent as contrasted with the presumably satisfactory but actually incorrect original naïve understanding. At one time an important game bird in Norway, the willow grouse, was so clearly threatened with extinction that it was thought wise to establish protective regulations and to place a bounty on its chief enemy, a hawk which was known to feed heavily on it. Quantities of the hawks were exterminated, but despite such drastic measures the grouse disappeared actually more rapidly than before. The naïvely applied customary remedies had obviously failed. But instead of becoming discouraged and quietistically letting this bird go the way of the great auk and the passenger pigeon, the authorities enlarged the scope of their investigations until the anomaly was explained. An ecological analysis into the relational aspects of the situation disclosed that a parasitic disease, coccidiosis, was epizootic among the grouse. In its incipient stages, this disease so reduced the flying speed of the grouse that the mildly ill individuals became easy prey for the hawks. In living largely off the slightly ill birds, the hawks prevented them from developing the disease in its full intensity and so spreading it more widely and quickly to otherwise healthy fowl. Thus the presumed enemies of the grouse, by controlling the epizootic aspects of the disease, proved to be friends in disguise.

In summarizing the above situation, the measure of validity wouldn't be to assume that, even in the well-understood factor of coccidiosis, we have the real "cause" of any beneficial or untoward condition, but to say, rather, that in this phase we have a highly significant and possibly preponderantly important relational aspect of the picture.

However, many people are unwilling to chance the sometimes ruthless-appearing notions which may arise through non-teleological treatments. They fear even to use them in that they may be left dangling out in space, deprived of such emotional support as had been afforded them by an unthinking belief in the proved value of pest control in the conservation of game birds; in the institutions of tradition; religion; science; in the security of the home or the

[1] Abstracted from the article on ecology by Elton, *Encyclopaedia Britannica,* 14th Edition, Vol. VII, p. 916.

family; or in a comfortable bank account. But for that matter emancipations in general are likely to be held in terror by those who may not yet have achieved them, but whose thresholds in those respects are becoming significantly low. Think of the fascinated horror, or at best tolerance, with which little girls regard their brothers who have dispensed with the Santa Claus belief; or the fear of the devout young churchman for his university senior who has grown away from depending on the security of religion.

As a matter of fact, whoever employs this type of thinking with other than a few close friends will be referred to as detached, hard-hearted, or even cruel. Quite the opposite seems to be true. Non-teleological methods more than any other seem capable of great tenderness, of an all-embracingness which is rare otherwise. Consider, for instance, the fact that, once a given situation is deeply understood, no apologies are required. There are ample difficulties even to understanding conditions "as is." Once that has been accomplished, the "why" of it (known now to be simply a relation, though probably a near and important one) seems no longer to be preponderantly important. It needn't be condoned or extenuated, it just "is." It is seen merely as part of a more or less dim whole picture. As an example: A woman near us in the Carmel woods was upset when her dog was poisoned—frightened at the thought of passing the night alone after years of companionship with the animal. She phoned to ask if, with our windows on that side of the house closed as they were normally, we could hear her ringing a dinner bell as a signal during the night that marauders had cut her phone wires preparatory to robbing her. Of course that was, in fact, an improbable contingency to be provided against; a man would call it a foolish fear, neurotic. And so it was. But one could say kindly, "We can hear the bell quite plainly, but if desirable we can adjust our sleeping arrangements so as to be able to come over there instantly in case you need us," without even stopping to consider whether or not the fear was foolish, or to be concerned about it if it were, correctly regarding all that as secondary. And if the woman had said apologetically, "Oh, you must forgive me; I know my fears are foolish, but I am so upset!" the wise reply would have been, "Dear person, nothing to forgive. If you have fears, they *are*; they are real things and to be considered. Whether or not they're foolish is beside the point. *What* they are is unimportant alongside the fact *that* they are." In other words, the badness or goodness, the teleology of the fears, was decidedly secondary. The whole notion could be conveyed by a smile or by a pleasant intonation more readily than by the words themselves. Teleological treatment which one might have been tempted to employ under the circumstances would first have stressed the fact that the fear was foolish—would say with a great show of objective justice, "Well, there's no use in *our* doing anything; the fault is that *your* fear is foolish and improbable. Get over that" (as a judge would say, "Come into court with clean hands"); "then if there's anything *sensible* we can do, we'll see," with smug blame implied in every word. Or, more kindly, it would try to reason with the woman in an attempt to help her get over it—the business of propaganda directed towards change even before the situation is fully understood (maybe as a lazy substitute for understanding). Or, still more kindly, the teleological method would try to understand the fear causally. But with the non-teleological treatment there is only the love and understanding of instant acceptance; after that fundamental has been achieved, the next step, if any should be necessary, can be considered more sensibly.

Strictly, the term non-teleological thinking ought not to be applied to what we have in mind. Because it involves more than thinking, that term is inadequate. *Modus operandi* might be better—a method of handling data of any sort. The example cited just above concerns feeling more than thinking. The method extends beyond thinking even to living itself; in fact, by inferred definition it transcends the realm of thinking possibilities, it postulates "living into."

In the destitute-unemployed illustration, thinking, as being the evaluatory function chiefly concerned, was the point of departure, "the crust to break through." There the "blame approach" considered the situation in the limited and inadequate teleological manner. The non-teleological method included that viewpoint as correct but limited. But when it came to the feeling aspects of a human relation situation, the non-teleological method would probably ameliorate the woman's fears in a loving, truly mellow, and adequate fashion, whereas the teleological would have tended to bungle things by employing the limited and sophisticated approach.

Incidentally, there is in this connection a remarkable etiological similarity to be noted between cause in thinking and blame in feeling. One feels that one's neighbors are to be blamed for their hate or anger or fear. One thinks that poor pavements are "caused" by politics. The non-teleological picture in either case is the larger one that goes beyond blame or cause. And the non-causal or non-blaming viewpoint seems to us very often relatively to represent the "new thing," the Hegelian "Christ-child" which arises emergently from the union of two opposing viewpoints, such as those of physical and spiritual teleologies, especially if there is conflict as to causation between the two or within either. The new viewpoint very frequently sheds light over a larger picture, providing a key which may unlock levels not accessible to either of the teleological viewpoints. There are interesting parallels here: to the triangle, to the Christian ideas of trinity, to Hegel's dialectic, and to Swedenborg's metaphysic of divine love (feeling) and divine wisdom (thinking).

The factors we have been considering as "answers" seem to be merely symbols or indices, relational aspects of things—of which they are integral parts—not to be considered in terms of causes and effects. The truest reason for anything's being so is that it *is*. This is actually and truly a reason, more valid and clearer than all the other separate reasons, or than any group of them short of the whole. Anything less than the whole forms part of the picture only, and the infinite whole is unknowable except by *being* it, by living into it.

A thing may be *so* "because" of a thousand and one reasons of greater or lesser importance, such as the man oversized because of glandular insufficiency. The integration of these many reasons which are in the nature of relations rather than reasons is that he *is*. The separate reasons, no matter how valid, are only fragmentary parts of the picture. And the whole necessarily includes all that it impinges on as object and subject, in ripples fading with distance or depending upon the original intensity of the vortex.

The frequent allusions to an underlying pattern have no implication of mysticism—except inasmuch as a pattern which comprises infinity in factors and symbols might be called mystic. But infinity as here used occurs also in the mathematical aspects of physiology and physics, both far away from mysticism as the term is ordinarily employed. Actually, the underlying pattern is probably nothing more than an integration of just such symbols and indices and mutual reference points as are already known, except that its power is *n*. Such an integration might include nothing more spectacular than we already know. But, equally, it *could* include anything, even events and entities as different from those already known as the vectors, tensors, scalars, and ideas of electrical charges in mathematical physics are different from the mechanical-model world of the Victorian scientists.

In such a pattern, causality would be merely a name for something that exists only in our partial and biased mental reconstructings. The pattern which it indexes, however, would be real, but not intellectually apperceivable because the pattern goes everywhere and is everything and cannot be encompassed by finite mind or by anything short of life—which it is.

The psychic or spiritual residua remaining after the most careful physical analyses, or the physical remnants obvious, particularly to us of the twentieth century, in the most honest and disciplined spiritual speculations of medieval philosophers, all bespeak such a pattern. Those residua, those most minute differentials, the 0.001 percentages which suffice to maintain the races of sea animals, are seen finally to be the most important things in the world, not because of their sizes, but because they are everywhere. The differential is the true universal, the true catalyst, the cosmic solvent. Any investigation carried far enough will bring to light these residua, or rather will leave them still unassailable as Emerson remarked a hundred years ago in "The Oversoul"—will run into the brick wall of the *impossibility* of perfection while at the same time insisting on the *validity* of perfection. Anomalies especially testify to that framework; they are the commonest intellectual vehicles for breaking through; all are solvable in the sense that any *one* is understandable, but that one leads with the power *n* to still more and deeper anomalies.

This deep underlying pattern inferred by non-teleological thinking crops up everywhere—a relational thing, surely, relating opposing factors on different levels, as reality and potential are related. But it must not be considered as causative, it simply exists, it *is*, things are merely expressions of it as it is expressions of them. And they *are* it, also. As Swinburne, extolling Hertha, the earth goddess, makes her say: "Man, equal and one with me, man that is made of me, man that is I," so all things which are *that*—which is all—equally may be extolled. That pattern materializes everywhere in the sense that Eddington finds the non-integer *q* "number" appearing everywhere, in the background of all fundamental equations,[2] in the sense that the speed of light, constant despite compoundings or subtractions, seemed at one time almost to be conspiring against investigation.

The whole is necessarily everything, the whole world of fact and fancy, body and psyche, physical fact and spiritual truth, individual and collective, life and death, macrocosm and microcosm (the greatest quanta here, the greatest synapse between these two), conscious and unconscious, subject and object. The whole picture is portrayed by *is,* the deepest word of deep ultimate reality, not shallow or partial as reasons are, but deeper and participating, possibly encompassing the Oriental concept of *being.*

And all this against the hot beach on an Easter Sunday, with the passing day and the passing time. This little trip of ours was becoming a thing and a dual thing, with collecting and eating and sleeping merging with the thinking-speculating activity. Quality of sunlight, blueness and smoothness of water, boat engines, and ourselves were all parts of a larger whole and we could begin to feel its nature but not its size.

[2] *The Nature of the Physical World,* pp. 208–10.

15

About noon we sailed and moved out of the shrouded and quiet Amortajada Bay and up the coast toward Marcial Reef, which was marked as our next collecting station. We arrived in mid-afternoon and collected on the late tide, on a northerly pile of boulders, part of the central reef. This was just south of Marcial Point, which marks the southern limit of Agua Verde Bay.

It was not a good collecting tide, although it should have been according to the tide chart. The water did not go low enough for exhaustive collecting. There were a few polyclads which here were high on the rocks. We found two large and many small chitons—the first time we had discovered them in numbers. There were many urchins visible but too deep below the surface to get to. Swarms of larval shrimps were in the water swimming about in small circles. The collecting was not successful in point of view of numbers of forms taken.

That night we rigged a lamp over the side, shaded it with a paper cone, and hung it close down to the water so that the light was reflected downward. Pelagic isopods and mysids immediately swarmed to the illuminated circle until the water seemed to heave and whirl with them. The small fish came to this horde of food, and on the outer edges of the light ring large fishes flashed in and out after the small fishes. Occasionally we interrupted this mad dance with dip-nets, dropping the catch into porcelain pans for closer study, and out of the nets came animals small or transparent that we had not noticed in the sea at all.

Having had no good tide at Marcial Reef, we arose at four o'clock the following morning and went in the darkness to collect again. We carried big seven-cell focusing flashlights. In some ways they make collecting in the dark, in a small area at least, more interesting than daytime collecting, for they limit the range of observation so that in the narrowed field one is likely to notice more detail. There is a second reason for our preference for night collecting—a number of animals are more active at night than in the daytime and they seem to be not much disturbed or frightened by artificial light. This time we had a very fair tide. The light fell on a monster highly colored spiny lobster in a crevice of the reef. He was blue and orange and spotted with brown. The taking of him required caution, for these big lobsters are very strong and are so armed with spikes and points that in struggling with one the hands can be badly cut. We approached with care, bent slowly down, and then with two hands grabbed him about the middle of the body. And there was no struggle whatever. He was either sick or lazy or hurt by the surf, and did not fight at all.

The cavities in Marcial Reef held a great many club-spined urchins and a number of the sharp-spined purple ones which had hurt us before. There were numbers of sea-fans, two of the usual starfish and a new species[1] which later we were to find common farther north in the Gulf. We took a good quantity of the many-rayed sun-stars, and a flat kind of cucumber which was new to us.[2] This was the first time we had collected at night, and under our lights we saw the puffer fish lazily feeding near the surface in the clear water. On the bottom, the brittle-stars, which we had always found under rocks, were crawling about like thousands of little snakes. They rarely move about in the daylight. Wherever the sharp, powerful rays of the flashlight cut into the water we could see the moving beautiful fish and the bottoms alive with busy feeding invertebrates. But collecting with a flashlight is difficult unless it is arranged that two people work together—one to hold the light and the other to take the animals. Also, from constant wetting in salt water the life of a flashlight is very short.

The one huge and beautiful lobster was the prize of this trip. We tried to photograph him on color film and as usual something went wrong but we got a very good likeness of one end of him, which was an improvement on our previous pictures. In most of our other photographs we didn't get either end.

We took several species of chitons and a great number of tunicates. There were several turbellarian flatworms, but these are so likely to dissolve before they preserve that we had great difficulties with them. There were in the collecting pans several species of brit-

[1] *Othilia tenuispinus.*

[2] Probably *Stichopus fuscus*—the specimen has since been lost sight of.

tle-stars, numbers of small crabs and snapping shrimps, plumularian hydroids, bivalves of a number of species, snails, and some small sea-urchins. There were worms, hermit crabs, sipunculids, and sponges. The pools too had been thick with pelagic larval shrimps, pelagic isopods—tiny crustacea similar to sow-bugs—and tiny shrimps (mysids). In this area the water seemed particularly peopled with small pelagic animals—"bugs," so the boys said. Everywhere there were bugs, flying, crawling, and swimming. The shallow and warm waters of the area promoted a competitive life that was astonishing.

After breakfast we pulled up the anchor and set out again northward. The pattern of the technique of the trip had by now established itself almost as a habit with us; collecting, running to a new station, collecting again. The water was intensely blue on this run, and the fish were very many. We could see the splashing of great schools of tuna in the distance where they beat the water to spray in their millions. The swordfish leaped all about us, and someone was on the bow the whole time trying to drive a light harpoon into one, but we never could get close enough. Cast after cast fell short.

We preserved and labeled as we went, and the water was so smooth that we had no difficulty with delicate animals. If the boat rolls, retractile animals such as anemones and sipunculids are more than likely to draw into themselves and refuse to relax under the Epsom-salts treatment, but this sea was as smooth as a lawn, and our wake fanned out for miles behind us.

The fish-lines on the stays snapped and jerked and we brought in skipjack, Sparky's friend of the curious name, and the Mexican sierra. This golden fish with brilliant blue spots is shaped like a trout. In size it ranges from fifteen inches to two feet, is slender and a very rapid swimmer. The sierra does not seem to travel in dense, surface-beating schools as the tuna does. Although it belongs with the mackerel-like forms, its meat is white and delicate and sweet. Simply fried in big hunks, it is the most delicious fish of all.

16

MARCH 25

About noon we arrived at Puerto Escondido, the Hidden Harbor, a place of magic. If one wished to design a secret personal bay, one would probably build something very like this little harbor. A point swings about, making a small semicircular bay fringed with bright-green mangroves, and only when one has turned inside this outer bay can one see that there is a second, secret bay beyond—a long narrow bay with an entrance not more than fifty feet wide at flood. The charts gave three fathoms at the center of the entrance, but the tide run was so furious that we did not attempt to take the *Western Flyer* in, but anchored in back of the first point, called Piedra de la Marina. Here we had more than ten fathoms, and Tony felt better about it.

In the distance, and from the south, a canoe came up the coast with a small sail set. The Indians move great distances in their tiny boats. As soon as the anchor was out, we dropped the fishing lines and immediately hooked several hammer-head sharks and a large red snapper. The air here was hot and filled with the smell of mangrove flowers. The little outer bay was our first collecting station, a shallow warm cove with a mud bottom and edged with small boulders, smooth and unencrusted with algae. On the bottom we could see long snake-like animals, gray with black markings, with purplish-orange floriate heads like chrysanthemums. They were about three feet long and new to us. Wading in rubber boots, we captured some of them and they proved to be giant synaptids.[1] They were strange and frightening to handle, for they stuck to anything they touched, not with slime but as though they were coated with innumerable suction-cells. On being taken from the water, they collapsed to skin, for their bodily shape is maintained by the current of water which they draw through themselves. When lifted out, this water escapes and they hang as limp as unfilled sausage

[1] A worm-like sea-cucumber, *Euapta godeffroyi.*

skins. Since they were new and fascinating to us, we took many specimens, maneuvering them gently to the surface and then sliding them into submerged wooden collecting buckets to prevent them from dropping their water. On the bottom they crawled about, their flower-heads moving gently, while the current of water passing through their bodies drew food into their stomachs. When we took them on board, we found they had to a high degree the habit of a number of holothurians: eviscerating. These *Euapta* were a nervous lot. We tried to relax them with Epsom salts so that we might kill them with their floriate heads extended, but the salts, no matter how carefully administered, caused the heads to retract, and soon afterwards they threw their stomachs out into the water. The word "stomach" is used here inadvisedly, for what they actually disgorge is the intestinal tract and respiratory tree.

We intoxicated them with pure oxygen and then tried the salts, but with the same result. Finally, by administering the salts in minute quantities and very slowly, we were able to preserve some uneviscerated specimens, but none with the head extended. The color motion pictures of the living animals, while not very good, at least showed the color and shape and movement of the extended heads. Again we got photographs of only one end, but this time the more important end, the floriate head.

In the little shallow bay there were many bright-green gars, or needle-fish, but they were too fast for our dip-nets and we were unable to take them. *Botete,* the poison fish, was here also in great numbers, and the boys took some of them with a light seine. We found here two new starfishes and many *Cerianthus* anemones.

While we were collecting on the shore, Tiny rowed about in the little skiff in slightly deeper water. He carried a light three-pronged spear with which he picked up an occasional cushion star from the bottom. We heard him shout, and looked up to see a giant manta ray headed for him, the tips of the wings more than ten feet apart. It was rare to see them in such shallow water. As it passed directly under his boat we yelled at him to spear it, since he wanted to so badly, but he simply sat in the bottom of the boat, gazing after the retreating ray, weakly swearing at us. For a long time he sat there quietly, not quite believing what he had seen. This great fish could have flicked Tiny and boat and all into the air with one flap of its wing. Tiny wanted to sit still and think for a long time and he did.

For an hour afterward he could only repeat, "Did you see that Goddamned thing!" And from that moment it became Tiny's ambition to catch and kill one of the giant rays.

The canoe which had been sailing up the coast came alongside and a man and a little boy boarded us. They had with them what they called "abalon"—not true abalones, but gigantic fixed scallops, very good for food. They had also some of the hacha, the huge fan-shaped clam; pearl oysters, which are growing rare; and several huge conchs. We bought from the man what he had and asked him to get us more of the large shellfish. We might look for weeks for animals he could go to directly. Everywhere it is the same: if an animal is good to eat or poisonous or dangerous the natives of the place will know about it and where it lives. But if it have none of these qualities, no matter how highly colored or beautiful, he may never in his life have seen it.

On the stone-bordered sandspit which is the southern block to the true inner Puerto Escondido there was a new stone building not quite finished, with no one about it. Around the point there now came a large rowboat pushed by a fast outboard motor of a species distinct from the Sea-Cow, for it seemed controlled and dominated by its master. In this boat there were several Indians and three men dressed in riding breeches and hiking boots. They came aboard and introduced themselves as Leopoldo Pérpuly, who owned a ranch on the edge of Puerto Escondido, Gilbert Baldibia, a school-teacher from Loreto, and Manuel Madinabeitia C., of the customs service, also of Loreto. These last two were on a vacation and hunting trip. They were strong, fine-looking men wearing the ever-present .45-caliber automatics of the government service. We served them canned fruit salad and discussed with them the country we had covered, and they asked us to go hunting the *borrego,* or big-horn sheep, with them, starting that afternoon and getting back the next day. We were to go into the tremendous and desolate stone mountains to camp and hunt. We accepted immediately, and went with them to the little ranch set back half a mile from Puerto Escondido. We didn't want to kill a big-horn sheep, but we wanted to see the country. As it turned out, none of them—the rancher, the teacher, or the customs man—had any intention of killing a big-horn sheep.

The little ranch was set deep in the brush. It was watered by deep wells of brackish brown water out of which endless chains of buckets emerged at the insistence of mules which turned the windlass. This rancher in the desert has dug sixty-foot wells, and he is raising tomatoes and he has planted many grapevines. But so dry is the earth that a few weeks without the rising buckets would destroy all his work. The houses of the ranch were simply roofs and low walls of woven palm, enough to keep out the wind but no obstruction to the air. The floors were of swept hard-packed earth, and there was an air of comfort about the place. The Indian workmen worked very slowly, and the babies peeked out of the woven houses at us. We were to ride to the mountains on mules and one small horse while two Indian men walked ahead. We were sorry for them until we discovered that their main irritation lay in the fact that horses and mules are so slow. Often they disappeared ahead of us, and we found them later sitting beside the trail waiting for us. The line of us started out on a clear but unfinished road that was eventually to go to Loreto. The thick and thorny brush and cactus had been grubbed, but no scraping had been done yet. It was a fantastic country; heavy xerophytic plants: cacti, mimosa, and thorned bushes and trees crackled with the heat. There were the lichens which bleed bright red when they are broken and were once a source of dye before the anilines were developed. There were poison bushes which we were warned about, for if one touches them and then rubs one's eyes, blindness ensues. We learned some of the uses of plants of this country; maidenhair fern, we were told, is boiled to an infusion and given to women after childbirth. It is said that no hemorrhage can follow this treatment. We rode over a rolling, rocky, desolate country, then left the cleared, some-day road and turned up a trail toward the stone mountains, steep and slippery with shale. And here our Indians were even more impatient, for the mules went more slowly while the Indians did not change gait for the steep places.

"My mule was a complainer. For a while I thought he simply didn't like me, but I believe now that he had a sour eye for the world. With every step he groaned with pain so convincingly that once I removed the saddle to see whether he might not be saddle-burned. He did not grunt, but drew from deep in his belly great groans of an agonized soul left to molder in Purgatory. It is impossible to see why he did this, for certainly no Mexican would believe him and he had never carried one of the more sentimental northern race before. I was heart-broken for him, but not sufficiently to get off and walk. We both suffered up the trail, he with pain and I with sorrow for him." (Extract from the personal journal of one of us.)

The trail cut back on itself again and again, and the bare mountains towered high and brooding over us. Far below we could see the brilliant blue water of the Gulf with a fantastic mirage cast over it.

There was in our party one horse, a spindle-legged, small-buttocked little animal with eyes haunted by social inadequacy; one horse in a society of mules, and a gelding at that. We thought how often one mule is surrounded by socially dominant horses, all grace and prance, conscious of their power and loveliness. In this pattern the mule has developed his anti-social self-sufficiency. He knows he can out-think a horse and he is pretty sure he can out-think a human. In both respects he is correct. And so your socially outcast mule dwells inward in sneering intellectuality; his mental pattern, conditioned by centuries of this cynical intellectualism, is set, and he is complete, sullen, treacherous, loving no one, selfish and self-centered. But this horse, having no such background, was unable to make the change in one generation. Surrounded by mules, he sorrowed and his spirit broke and his eyes were sad. The stiffening was gone from his ears and his mouth hung open. He slunk ashamedly along behind the mules. Stripped of his regalia and his titles, he was a pitiful thing. Refugee princes usually become waiters, but this poor horse was not even able to be a waiter, let alone a horse. And just as one is irritated by a grand duke if he has no robes and garters and large metal-and-enamel decorations, so we found ourselves disliking this poor horse; and he knew it and it didn't help him.

We came at last to a trail of broken stone and rubble so steep that the mules could not carry us any more. We dismounted and crawled on all fours, and we don't know how the mules got up. After a short climb we emerged on a level place in a deep cleft in the granite mountains. In this cleft a tiny stream of water fell hundreds of feet from pool to pool. There were palm trees and wild grapevines and large ferns, and the water was cool and sweet. This little stream, coming from so high up in the mountains and falling so far, never had the final dignity of reaching the ocean. The desert sucked it down and the heat dried it up and on the level it disappeared in a light mist of frustration. We sat beside a pool of the waterfall and our Indians made coffee for us and unpacked a lunch, and one item of this lunch was so delicious that we have wanted it again. It is made in this way: a warm tortilla is laid down and spread with well-cooked beans, and another tortilla laid on top and spread, and another, until it is ten or twelve layers thick. Then it is wrapped in cloth. Before eating it one slices downward through the layers as with a cake. It is a fine dish and very filling. While we ate, the Indians made our beds on the ground, and we fired a few shots at a rock across the canyon. Then it was dark and we lay in our blankets and talked, and here we suffered greatly. For the funny stories began. We suppose they weren't clean stories, but we couldn't be sure. Nearly every one began, "Once there was a school-teacher with large black eyes—very sympathetic—" "*Muy simpática*" has a slightly different connotation from that of "sympathetic," for sympathy is a passive state of receptivity, but to be "*simpática*" is to be more active or co-operative, even sometimes a little forward. At any rate, this "*simpática*" school-teacher invariably had as one of her students "a tall strong boy, *con cojones, pero cojones*"—this last with a gesture easily seen in the firelight. The stories progressed until they came to the snappers; we leaned forward studiously intent, but the snappers were either so colloquial that we could not understand them or so filled with the laughter of the teller that we couldn't make out the words. Story after story was told, and we didn't get a single snapper, not one. Our suspicions were aroused of course. We knew something was bound to happen when a school-teacher "*muy simpática*" asks a large boy "*con cojones*" to stay after school, but whether it ever did or not we do not know.

It grew cold in the night, and the mosquitoes were unmerciful. In this sparsely populated country human blood must be a rarity. We were a seldom-found dessert to them, and they whooped and screamed and attacked, power-diving and wheeling up and diving again. The visibility was good, and we made excellent targets. Only when it became bitterly cold did they go away.

We have noticed many times how lightly Mexican Indians sleep. Often in the night they awaken to smoke a cigarette and talk softly together for a while, and then go to sleep again rather like restless birds, which sing a little in the dark, dreaming that it is already day. Half a dozen times a night they may awaken thus, and it is pleasant to hear them, for they talk very quietly as though they were dreaming.

When the dawn came, our Indians made coffee for us and we ate more of the lunch. Then, with some ceremony, the ranch-owner presented a Winchester .30-30 carbine with a broken stock to those Indians, and they set off straight up the mountainside. This, our first hunt for the *borrego,* or big-horn sheep, was the nicest hunting we have ever had. We did not raise a hand in our own service during the entire trip. Besides, we do not like to kill things—we do it when it is necessary, but we take no pleasure in it; and those fine Indians did it for us—the hunting, that is—while we sat beside the little waterfall and discussed many things with our hosts—how all Americans are rich and own new Fords; how there is no poverty in the United States and everyone sees a moving picture every night and is drunk as often as he wishes; how there are no political animosities; no need; no fear; no failure; no unemployment or hunger. It was a wonderful country we came from and our hosts knew all about it and told us. We could not spoil such a dream. After each one of his assurances we said, "*Cómo no?*" which is the most cautious understatement in the world, for "*Cómo no?*" means nothing at all. It is a polite filler between two statements from your companion. And we sat in that cool place and looked out over the hot desert country to the blue Gulf. In a couple of hours our Indians came back; they had no *borrego,* but one of them had a pocketful of droppings. It was time by now to start back to the boat. We intend to do all our future hunting in exactly this way. The ranch-owner said a little sadly, "If they had killed one we could have had our pictures taken with it," but except for that loss, there was no loss, for none of us likes to have the horns of dead animals around.

We had sat beside the little pool and watched the tree-frogs and the horsehair worms and the water-skaters, and had wondered how they got there, so far from other water. It seemed to us that life in every form is incipiently everywhere waiting for a chance to take root and start reproducing; eggs, spores, seeds, bacilli—everywhere. Let a raindrop fall and it is crowded with the waiting life. Everything is everywhere; and we, seeing the desert country, the hot waterless expanse, and knowing how far away the nearest water must be, say with a kind of disbelief, "How did they get clear here, these little animals?" And until we can attack with our poor blunt weapon of reason that causal process and reduce it, we do not quite believe in the horsehair worms and the tree-frogs. The great fact is that they are there. Seeing a school of fish lying quietly in still water, all the heads pointing in one direction, one says, "It is unusual that this is so"—but it isn't unusual at all. We begin at the wrong end. They simply lie that way, and it is remarkable only because with our blunt tool we cannot carve out a human reason. Everything is potentially everywhere—the body is potentially cancerous, phthisic, strong to resist or weak to receive. In one swing of the balance the waiting life pounces in and takes possession and grows strong while our own individual chemistry is distorted past the point where it can maintain its balance. This we call dying, and by the process we do not give nor offer but are taken by a multiform life and used for its proliferation. These things are balanced. A man is potentially all things too, greedy and cruel, capable of great love or great hatred, of balanced or unbalanced so-called emotions. This is the way he is—one factor in a surge of striving. And he continues to ask "why" without first admitting to himself his cosmic identity. There are colonies of pelagic tunicates[2] which have taken a shape like the finger of a glove. Each member of the colony is an individual animal, but the colony is another individual animal, not at all like the sum of its individuals. Some of the colonists, girdling the open end, have developed the ability, one against the other, of making a pulsing movement very like muscular action. Others of the colonists collect the food and distribute it, and the outside of the glove is hardened and protected against contact. Here are two animals, and yet the same thing—something the early Church would have been forced to call a mystery. When the early Church called some matter "a mystery" it accepted that thing fully and deeply as *so*, but simply not accessible to reason because reason had no business with it. So a man of individualistic reason, if he must ask, "Which is the animal, the colony or the individual?" must abandon his particular kind of reason and say, "Why, it's two animals and they aren't alike any more than the cells of my body are

[2] *Pyrosoma giganteum.*

like me. I am much more than the sum of my cells and, for all I know, they are much more than the division of me." There is no quietism in such acceptance, but rather the basis for a far deeper understanding of us and our world. And now this is ready for the taboo-box.

It is not enough to say that we cannot know or judge because all the information is not in. The process of gathering knowledge does not lead to knowing. A child's world spreads only a little beyond his understanding while that of a great scientist thrusts outward immeasurably. An answer is invariably the parent of a great family of new questions. So we draw worlds and fit them like tracings against the world about us, and crumple them when they do not fit and draw new ones. The tree-frog in the high pool in the mountain cleft, had he been endowed with human reason, on finding a cigarette butt in the water might have said, "Here is an impossibility. There is no tobacco hereabouts nor any paper. Here is evidence of fire and there has been no fire. This thing cannot fly nor crawl nor blow in the wind. In fact, this thing cannot be and I will deny it, for if I admit that this thing is here the whole world of frogs is in danger, and from there it is only one step to anti-frogicentricism." And so that frog will for the rest of his life try to forget that something that is, is.

On the way back from the mountain one of the Indians offered us his pocketful of sheep droppings, and we accepted only a few because he did not have many and he probably had relatives who wanted them. We came back through heat and dryness to Puerto Escondido, and it seemed ridiculous to us that the *Western Flyer* had been there all the time. Our hosts had been kind to us and considerate as only Mexicans can be. Furthermore, they had taught us the best of all ways to go hunting, and we shall never use any other. We have, however, made one slight improvement on their method: we shall not take a gun, thereby obviating the last remote possibility of having the hunt cluttered up with game. We have never understood why men mount the heads of animals and hang them up to look down on their conquerors. Possibly it feels good to these men to be superior to animals, but it does seem that if they were sure of it they would not have to prove it. Often a man who is afraid must constantly demonstrate his courage and, in the case of the hunter, must keep a tangible record of his courage. For ourselves, we have had mounted in a small hardwood plaque one perfect *borrego* dropping. And where another man can say, "There was an animal, but because I am greater than he, he is dead and I am alive, and there is his head to prove it," we can say, "There was an animal, and for all we know there still is and here is the proof of it. He was very healthy when we last heard of him."

After the dryness of the mountain it was good to come back to the sea again. One who was born by the ocean or has associated with it cannot ever be quite content away from it for very long.

Sparky made us a great dish of his spaghetti, the veritable Enea spaghetti, and we ate until we were bloated with it.

Now our equipment began to show its weaknesses. The valve of the oxygen cylinder gave trouble owing to the humidity. The little ice-plant was not powerful enough, and where it should have cooled sea water for us, it was all it could do to keep the beer chilled. Besides, it broke down very often.

By now, some animals began to emerge as ubiquitous. *Heliaster kubiniji*, the sun-star, was virtually everywhere, but we did observe that the farther up the Gulf we went, the smaller he became. *Eurythoë*, the stinging worm, occurred wherever there were loosely imbedded rocks or coral under which he could hide. In this connection it is interesting that in the description of this worm in Chamberlin,[3] the one descriptive item completely ignored is the one most important to the collector—that he stings like the devil, his hair-like fringe breaking off in the hands and leaving a burn which does not disappear for a long time. Tiny, who is able to translate experience readily into emotion, found that anger did not overcome *Eurythoë*, and he grew to have the greatest respect for the worm, even to the point of adopting the usual collector's caution of never putting the hands where one hasn't looked first.

The purple sharp-spined urchin[4] occurred wherever there was rock or reef exposed to wave-shock or fast-scouring currents. There were the usual barnacles and limpets on the rocks high up in the littoral wherever their pattern of alternating water and air was available. Anemones, the small bunodid forms, were everywhere too. And, of course, the porcelain crabs, hermit crabs, and sea-cucumbers.

[3] "The Annelida Polychaeta," 1919, p. 28.

[4] *Arbacia incisa.*

We had taken a great many animals and, as compared with the work of some expensive, well-equipped, well-manned expeditions, our results began to cause us to wonder what methods were used by those collectors. For instance, the best reports to date (with the possible exception of the Hancock Expedition reports—and these are so expensive and rare that an amateur cannot afford them, and even university libraries do not always have them) are those of a well-known scientific expedition into the Gulf, about thirty years ago. There were eight naturalists aboard a specially built and equipped steamboat, with a complete and well-trained crew. In two months out of San Francisco they occupied thirty-five stations and took a total of 2351 individuals of 118 species of echinoderms both from deep water (including dredge hauls down to 1760 fathoms) and from along shore, and in two great faunal provinces. Only 39 species were from shallow water; 31 of these, in about 387 individuals, were from the Gulf. Already, in only nine days of Gulf collecting, in the one zoogeographical province and entirely along shore, we had taken almost double their 31 Gulf echinoderm species—the only group we had so far tabulated—and had begun to restrain our enthusiasm owing to the lack of containers. We worked hard, but not beyond reason, and our wonder is caused not by the numbers we took, but by the small numbers they did. We had time to play and to talk, and even to drink a little beer. (We took 2160 individuals of two species of beer.)

The shores of the Gulf, so rich for the collector, must still be fairly untouched (again except for the largely unreported Hancock collections). We had not the time for the long careful collecting which is necessary before the true picture of the background of life can be established. We rushed through because it was all we could afford, but our results seem to indicate that energy and enthusiasm can offset lack of equipment and personnel.

17

MARCH 27

We had collected extensively on the outer parts of Puerto Escondido, but not in the inner bay itself. At five-thirty A.M. Mexican time, we set out to circle this inner bay in the little skiff. It was dark when we started, and we used the big flashlights for collecting. There was a good low tide, and we moved slowly along the shore, one rowing while the other inspected the bottom with the light. There was no ripple to distort the surface. The eastern shore was dominated by the big, flat, chocolate-brown holothurian.[1] They moved slowly along, feeding on the bottom, many hundreds of them. They far overshadowed in number any other animals in this area. There were many of the ruffled clams[2] with hard, thick, wavy shells. The under-rock fauna was not very rich. The eastern and northern shores were littered with shattered rock, recently enough splintered so that the edges were still sharp, and in this quiet bay no waves would have ground the edges smooth. Mangroves bordered a great part of the bay, and the spicy smell of their flowers was strong and pleasant. A few of the giant, snake-like synaptids that we had taken in the outer bay waved and moved on the bottom. As we rounded toward the westerly side of the bay, we came to sand flats and a change of fauna, for the big brown cucumbers did not live here. The dawn came as we moved along the sand flats. Two animals were at the waterside, about as large as small collies, dark brown, with a cat-like walk. In the half-light we could not see them clearly, and as we came nearer they melted away through the mangroves. Possibly they were something like giant civet-cats. They had undoubtedly been fishing at the water's edge. On the smooth sand bottom of this area there were clusters of knobbed, green coral (probably *Porites porosa*—no samples were gathered), but except for *Cerianthus* and a few bivalves this bottom was comparatively sterile.

[1] *Stichopus fuscus.*
[2] *Carditamera affinis.*

Rounding the southern end of the bay, we came again to the single narrow entrance where the water was rushing in on the returning tide, and here, suddenly, the area was incredibly rich in fauna. Here, where the water rushes in and out, bringing with it food and freshness, there was a remarkable gathering. Beautiful red and green cushion stars littered the rocky bottom. We found clusters of a solitary soft coral-like form[3] in great knobs and heads in one restricted location on the rocks. Caught against the rocks by the current was a very large pelagic coelenterate, in appearance like an anemone with long orange-pink tentacles, apparently not retractable. On picking him up we were badly stung. His nettle-cells were vicious, stinging even through the calluses of the palms, and hurting like a great many bee-stings. At this entrance also we took several giant sea-hares,[4] a number of clams, and one small specimen of the clam-like hacha. For hours afterwards the sting of the anemone remained. So very many things are poisonous and hurtful in these Gulf waters: urchins, sting-rays, morays, heart-urchins, this beastly anemone, and many more. One becomes very timid after a while. Barnacle-cuts, which are impossible to avoid, cause irritating sores. The fingers and palms become cross-hatched with cuts, and then very quickly, possibly owing to the constant soaking in salt water and the regular lifting of rocks, the hands become covered with a hard, almost horny, callus.

The Puerto Escondido station was one of the richest we visited, for it combined many kinds of environment in a very small area; sand bottom, stone shore, boulders, broken rock, coral, still, warm, shallow places, and racing tide. It is highly probable that careful and extended collecting would show that individuals of species of a very respectable proportion of the total Panamic fauna could be found in this tiny world. Barring surf-battered reef, every probable environment occurs within these few acres—a textbook exhibit for ecologists.

We took rock isopods, sponges, tunicates, turbellarians, chitons, bivalves, snails, hermit crabs, and many other crabs, Heteronereids and mysids pelagic at night, small ophiurans, limpets, and worms

[3] In superficial appearance it was identical with the figures of the West Indies *Zoanthus pulchellus* illustrated in Duerden's "Actinians of Porto Rico," 1902, *U. S. Fish Comm. Bulletin* for 1900, Vol. 2, pp. 321–74.

[4] *Dolabella californica.*

and even listed in our collecting notes for the day the horsehair worms from the little waterfall in the mountains.[5] We took six to eight species of cucumbers and eleven of starfish at this one station.

When we came back from the early morning collecting we sailed immediately for the port of Loreto. We were eager to see this town, for it was the first successful settlement on the Peninsula, and its church is the oldest mission of all. Here the inhospitability of Lower California had finally been conquered and a colony had taken root in the face of hunger and mishap. From the sea, the town was buried in a grove of palms and greenery. We dropped anchor and searched the shore with our glasses. A line of canoes lay on the beach and a group of men sat on the sand by the canoes and watched us; comfortable, lazy-looking men in white clothes. When our anchor dropped they got up and made for the town. Of course, they had to find their uniforms, and since Loreto was not very often visited and since the Governor had *not* recently been there, this may not have been so easy. There may have been some scurrying of errand-bound children from house to house, looking for tunics or belts or borrowing clean shirts. Señor the official had to shave and scent himself and dress. It all takes time, and the boat in the harbor will wait. It didn't look like much of a boat anyway, but at least it was a boat.

One fine thing about Mexican officials is that they greet a fishing boat with the same serious ceremony they would afford the *Queen Mary,* and the *Queen Mary* would have to wait just as long. This made us feel very good and not rebellious about the port fees—absent in this case! We came to them and they made us feel, not like stodgy people in a purse-seiner but like ambassadors from Ultra-Marina bringing letters of greeting out of the distances. It is no wonder that we too scurried for clean shirts, that Tony put on his master's cap, and Tiny polished the naval insignia on his, which he had come by no doubt honorably in a washroom in San Diego. We were not smart, not very alert, but we were clean and we smelled rather delicious. Sparky sprinkled us with shaving lotion and we filled the air with an odor of flowers. If the *brazo,* the double

[5] *Chorodes* sp., probably *C. occidentalis* Montgomery, according to J. T. Lucker of the U. S. National Museum, their No. 159124.

embrace, should be indicated by any feeling of uncontrollable good-will, we were ready.

The men came back to the beach in their uniforms, paddled out, and we passed the ceremony of induction. Loreto was asleep in the sunshine, a lovely town, with gardens in every yard and only the streets white and hot. The young males watched us from the safe shade of the *cantina* and passed greetings as we went by, and a covey of young girls grew tight-faced and rushed around a corner and giggled. How strange we were in Loreto! Our trousers were dark, not white; the silly caps we wore were so outlandish that no store in Loreto would think of stocking them. We were neither soldiers nor sailors—the little girls just couldn't take it. We could hear their strangled giggling from around the corner. Now and then they peeked back around the corner to verify for themselves our ridiculousness, and then giggled again while their elders hissed in disapproval. And one woman standing in a lovely garden shaded with purple bougainvillaea explained, "Everyone knows what silly things girls are. You must forgive their ill manners; they will be ashamed later on." But we felt that the silly girls had something worthwhile in their attitude. They were definitely amused. It is often so, particularly in our country, that the first reaction to strangeness is fear and hatred; we much preferred the laughter. We don't think it was even unkind—they'd simply never seen anything so funny in their lives.

As usual, a good serious small boy attached himself to us. It would be interesting to see whether a nation governed by the small boys of Mexico would not be a better, happier nation than those ruled by old men whose prejudices may or may not be conditioned by ulcerous stomachs and perhaps a little drying up of the stream of love.

This small boy could have been an ambassador to almost any country in the world. His straight-seeing dark eyes were courteous, yet firm. He was kind and dignified. He told us something of Loreto; of its poverty, and how its church was tumbled down now; and he walked with us to the destroyed mission. The roof had fallen in and the main body of the church was a mass of rubble. From the walls hung the shreds of old paintings. But the bell-tower was intact, and we wormed our way deviously up to look at the old bells and to strike them softly with the palms of our hands so that they glowed a little with tone. From here we could look down on the low roofs and into the enclosed gardens of the town. The white sunlight could not get into the gardens and a sleepy shade lay in them.

One small chapel was intact in the church, but the door to it was barred by a wooden grille, and we had to peer through into the small, dark, cool room. There were paintings on the walls, one of which we wanted badly to see more closely, for it looked very much like an El Greco, and probably was *not* painted by El Greco. Still, strange things have found their way here. The bells on the tower were the special present of the Spanish throne to this very loyal city. But it would be good to see this picture more closely. The Virgin Herself, Our Lady of Loreto, was in a glass case and surrounded by the lilies of the recently past Easter. In the dim light of the chapel she seemed very lovely. Perhaps she is gaudy; she has not the look of smug virginity so many have—the "I-am-the-Mother-of-Christ" look—but rather there was a look of terror in her face, of the Virgin Mother of the world and the prayers of so very many people heavy on her.

To the people of Loreto, and particularly to the Indians of the outland, she must be the loveliest thing in the world. It doesn't matter that our eyes, critical and thin with *good taste*, should find her gaudy. And actually we did not. We too found her lovely in her dim chapel with the lilies of Easter around her. This is a very holy place, and to question it is to question a fact as established as the tide. How easily and quickly we slide into our race-pattern unless we keep intact the stiff-necked and blinded pattern of the recent intellectual training.

We threw it over, and there wasn't much to throw over, and we felt good about it. This Lady, of plaster and wood and paint, is one of the strong ecological factors of the town of Loreto, and not to know her and her strength is to fail to know Loreto. One could not ignore a granite monolith in the path of the waves. Such a rock, breaking the rushing waters, would have an effect on animal distribution radiating in circles like a dropped stone in a pool. So has this plaster Lady a powerful effect on the deep black water of the human spirit. She may disappear and her name be lost, as the Magna Mater, as Isis, have disappeared. But something very like her will take her place, and the longings which created her will find somewhere in the world a similar altar on which to pour their force. No matter what her name is, Artemis, or Venus, or a girl behind a Woolworth counter vaguely remembered, she is as eternal as our species, and we will continue to manufacture her as long as we survive.

We came back slowly through the deserted streets of Loreto, and we walked quietly laden with submergence in a dim chapel.

A few supplies went aboard, and we pulled up the anchor and moved northward again. On the way we caught a Mexican sierra and another fish, apparently a cross between a yellow-fin tuna[6] and an albacore.[7] Tiny and Sparky, who have fished in tuna water a good deal, say that this cross is often found and taken, although never in numbers.

We sailed north and found anchorage on the northern end of Coronado Island, and went immediately to collect on a long, westerly-extending point. This reef of water-covered stones was not very rich. In high boots we moved slowly about, turning over the flattened algae-covered boulders. We found here many solitary corals,[8] and with great difficulty took some of them. They are very hard, and shatter easily when they are removed. If one could saw out the small section of rock to which they are fastened, it would be easy to take them. The next best method is to use a thin, very sharp knife and, by treating them as delicately as jewels, to remove them from their hard anchorage. Even with care, only about one in five is unbroken. Here also we found clustered heads of hard zoanthidean anemones of two types, one much larger than the other. We found a great number of large hemispherical yellow sponges which were noted in the collecting reports as "strikingly similar to the Monterey Bay *Tethya aurantia* or *Geodia*"—a similarity partly explainable by the fact that they turned out to be *Tethya aurantia* and a species of *Geodia!* Our collecting included the usual assortment of creatures, ranging from the crabs which plant algae on their backs for protection to the bryozoa which look more like moss than animals. With all these, the region was still

[6] *Neothunnus macropterus.*
[7] *Germo alalunga.*
[8] *Astrangia pederseni.*

not rich, but "burned," and again we felt the thing which had been at the strange Cayo Islet, a resentment of the shore toward animal life, an inhospitable quality in the stones that would make an animal think twice about living there.

It is so strange, this burned quality. We have seen places which seemed hostile to human life, too. There are parts of the coast of California which do not like humans. It is as though they were already inhabited by another and invisible species which resented humans. Perhaps such places are "burned" for us; perhaps a petrologist could say why. Might there not be a mild radio-activity which made one nervous in such a place so that he would say, trying to put words to his feeling, "This place is unfriendly. There is something here that will not tolerate my kind"? While some radio-activities have been shown to encourage not only life but mutation (note experimentation with fruit-flies), there might well be some other combinations which have an opposite effect.

Little fragments of seemingly unrelated information will sometimes accumulate in a process of speculation until a tenable hypothesis emerges. We had come on a riddle in our reading about the Gulf and now we were able to see this riddle in terms of the animals. There is an observable geographic differential in the fauna of the Gulf of California. The Cape San Lucas-La Paz area is strongly Panamic. Many warm-water mollusks and crustaceans are not known to occur in numbers north of La Paz, and some not even north of Cape San Lucas. But the region north of Santa Rosalia, and even of Puerto Escondido, is known to be inhabited by many colder-water animals, including *Pachygrapsus crassipes,* the commonest California shore crab, which ranges north as far as Oregon. These animals are apparently trapped in a blind alley with no members of their kind to the south of them.

The problem is: "How did they get there?" In 1895 Cooper[9] advanced an explanation. He remarks, referring to the northern part of the Gulf: "It appears that the species found there are more largely of the temperate fauna, many of them being identical with those of the same latitude on the west [outer] coast of the Peninsula.

[9] "Catalogue of Marine Shells . . . on Eastern Shore of Lower California . . . ," *Proc. Calif. Acad. Sci.,* Vol. 5 (2), p. 37.

This seems to indicate that the dividing ridge, now three thousand feet or more in altitude, was crossed by one or more channels within geologically recent times."

This differential, which we ourselves saw, has been remarked a number of times in the literature of the region, especially by conchologists. Eric Knight Jordan, son of David Starr Jordan, an extremely promising young paleontologist who was killed some years ago, studied the geological and present distribution of mollusks along the west coast of Lower California. He says[10]: "Two distinct faunas exist on the west coast of Lower California. The southern Californian *now* ranges southward from Point Conception to Cedros Island . . . probably extends a little farther. . . . The fauna of the Gulf of California ranges to the north on the west coast of the Peninsula approximately to Scammon's Lagoon, which is a little farther up than Cedros Island." Present geographical ranges are given for 124 species, collected in lower Quaternary beds at Magdalena Bay, all of which occur living today, but farther to the north. Two pages later he remarks: "It . . . appears that when these Quaternary beds were laid down there was a southward displacement of the isotherms sufficient to carry the conditions today prevailing at Cedros down as far as the latitude of Magdalena Bay."

Having reviewed the literature, we can confirm the significance of the Cedros Island complex as a present critical horizon (as Carpenter did eighty years ago) where the north and south fauna to some extent intermingle. Apparently this is the very condition that obtained at Magdalena Bay or southward when the lower Quaternary beds were being laid down. The present Magdalena Plain, extending to La Paz on the Gulf side, was at that time submerged. Then it was cold enough to permit a commingling of cold-water and warm-water species at that point. The hypothesis is tenable that when the isotherms retreated northward, the cold-water forms were no longer able to inhabit southern Lower California shores, which included the then Gulf entrance. In these increasingly warm waters they would have perished or would have been pushed northward, both along the outside coast, where they could retreat indefinitely, and into the Gulf. In the latter case the migrating waves of com-

[10] "Quaternary and Recent Molluscan Faunas of the West Coast of Lower California," *Bull. South Calif. Acad. Sci.*, Vol. 23 (5), p. 146.

peting animals from the south, which were invading the Gulf and spilling upward, would have pocketed the northern species in the upper reaches, where they have remained to this day. These animals, hemmed in by tropical waters and fortunate competitors, have maintained themselves for thousands of years, though in the struggle they have been modified toward pauperization.

This hypothesis would seem to offset Cooper's assumption of a channel through ridges some 350 miles to the north which show no signs of Quaternary submergence.

It is interesting that a paleontologist, working in one area, should lay the groundwork for a very reasonable hypothesis concerning the distribution of animals in another. It is, however, only one example among many of the obliqueness of investigation and the accident quotient involved in much investigation. The literature of science is filled with answers found when the question propounded had an entirely different direction and end.

There is one great difficulty with a good hypothesis. When it is completed and rounded, the corners smooth and the content cohesive and coherent, it is likely to become a thing in itself, a work of art. It is then like a finished sonnet or a painting completed. One hates to disturb it. Even if subsequent information should shoot a hole in it, one hates to tear it down because it once was beautiful and whole. One of our leading scientists, having reasoned a reef in the Pacific, was unable for a long time to reconcile the lack of a reef, indicated by soundings, with the reef his mind told him was there. A parallel occurred some years ago. A learned institution sent an expedition southward, one of whose many projects was to establish whether or not the sea-otter was extinct. In due time it returned with the information that the sea-otter was indeed extinct. One of us, some time later, talking with a woman on the coast below Monterey, was astonished to hear her describe animals living in the surf which could only be sea-otters, since she described accurately animals she couldn't have known about except by observation. A report of this to the institution in question elicited no response. It had extincted sea-otters and that was that. It was only when a reporter on one of our more disreputable newspapers photographed the animals that the public was informed. It is not yet known whether the institution of learning has been won over.

This is not set down in criticism; it is no light matter to make up one's mind about anything, even about sea-otters, and once made up, it is even harder to abandon the position. When a hypothesis is deeply accepted it becomes a growth which only a kind of surgery can amputate. Thus, beliefs persist long after their factual bases have been removed, and practices based on beliefs are often carried on even when the beliefs which stimulated them have been forgotten. The practice must follow the belief. It is often considered, particularly by reformers and legislators, that law is a stimulant to action or an inhibitor of action, when actually the reverse is true. Successful law is simply the publication of the practice of the majority of units of a society, and by it the inevitable variable units are either driven to conform or are eliminated. We have had many examples of law trying to be the well-spring of action; our prohibition law showed how completely fallacious that theory is.

The things of our minds have for us a greater toughness than external reality. One of us has a beard, and one night when this one was standing wheel-watch, the others sat in the galley drinking coffee. We were discussing werewolves and their almost universal occurrence in regional literature. From this beginning, we played with a macabre thought, "The moon will soon be full," we said, "and he of the beard will begin to feel the pull of the moon. Last night," we said, "we heard the scratch of claws on the deck. When you see him go down on all fours, when you see the red light come into his eyes, then look out, for he will slash your throat." We were delighted with the game. We developed the bearded one's tendencies, how his teeth, the canines at least, had been noticeably longer of late, how for the past week he had torn his dinner apart with his teeth. It was night as we talked thus, and the deck was dark and the wind was blowing. Suddenly he appeared in the doorway, his beard and hair blown, his eyes red from the wind. Climbing the two steps up from the galley, he seemed to arise from all fours, and everyone of us started, and felt the prickle of erecting hairs. We had actually talked and thought ourselves into this pattern, and it took a while for it to wear off.

These mind things are very strong; in some, so strong as to blot out the external things completely.

18

MARCH 28

After the collecting on Coronado Island, on the twenty-seventh, and the preservation and labeling, we found that we were very tired. We had worked constantly. On the morning of the twenty-eighth we slept. It was a good thing, we told ourselves; the eyes grow weary with looking at new things; sleeping late, we said, has its genuine therapeutic value; we would be better for it, would be able to work more effectively. We have little doubt that all this was true, but we wish we could build as good a rationalization every time we are lazy. For in some beastly way this fine laziness has got itself a bad name. It is easy to see how it might have come into disrepute, if the result of laziness were hunger. But it rarely is. Hunger makes laziness impossible. It has even become sinful to be lazy. We wonder why. One could argue, particularly if one had a gift for laziness, that it is a relaxation pregnant of activity, a sense of rest from which directed effort may arise, whereas most busy-ness is merely a kind of nervous tic. We know a lady who is obsessed with the idea of ashes in an ashtray. She is not lazy. She spends a good half of her waking time making sure that no ashes remain in any ashtray, and to make sure of keeping busy she has a great many ashtrays. Another acquaintance, a man, straightens rugs and pictures and arranges books and magazines in neat piles. He is not lazy, either; he is very busy. To what end? If he should relax, perhaps with his feet up on a chair and a glass of cool beer beside him—not cold, but cool—if he should examine from this position a rumpled rug or a crooked picture, saying to himself between sips of beer (preferably Carta Blanca beer), "This rug irritates me for some reason. If it were straight, I should be comfortable; but there is only one straight position (and this is of course, only my own personal discipline of straightness) among all possible positions. I am, in effect, trying to impose my will, my insular sense of rightness, on a rug, which of itself can have no such sense, since it seems equally contented straight or crooked. Suppose I should try to straighten people," and here he sips deeply. "Helen C., for instance, is not neat, and Helen C."—here he goes into a reverie—"how beautiful she is with her hair messy, how lovely when she is excited and breathing through her mouth." Again he raises his glass, and in a few minutes he picks up the telephone. He is happy; Helen C. may be happy; and the rug is not disturbed at all.

How can such a process have become a shame and a sin? Only in laziness can one achieve a state of contemplation which is a balancing of values, a weighing of oneself against the world and the world against itself. A busy man cannot find time for such balancing. We do not think a lazy man can commit murders, nor great thefts, nor lead a mob. He would be more likely to think about it and laugh. And a nation of lazy contemplative men would be incapable of fighting a war unless their very laziness were attacked. Wars are the activities of busy-ness.

With such a background of reasoning, we slept until nine A.M. And then the engines started and we moved toward Concepción Bay. The sea, with the exception of one blow outside of La Paz, had been very calm. This day, a little wind blew over the ultramarine water. The swordfish in great numbers jumped and played about us. We set up our lightest harpoon on the bow with a coil of cotton line beside it, and for hours we stood watch. The helmsman changed course again and again to try to bring the bow over a resting fish, but they seemed to wait until we were barely within throwing range and then they sounded so quickly that they seemed to snap from view. We made many wild casts and once we got the iron in, near the tail of a monster. But he flicked his tail and tore it out and was gone. We could see schools of leaping tuna all about us, and whenever we crossed the path of a school, our lines jumped and snapped under the strikes, and we brought the beautiful fish in.

We had set up a salt barrel near the stern, and we cut the fish into pieces and put them into brine to take home. It developed after we got home that several of us had added salt to the brine and the whole barrel was hopelessly salty and inedible.

As we turned Aguja Point and headed southward into the deep pocket of Concepción Bay, we could see Mulege on the northern shore—a small town in a blistering country. We had no plan for stopping there, for the story is that the port charges are mischievous and ruinous. We do not know that this is so, but it is repeated about Mulege very often. Also, there may be malaria there. We had been following the trail of malaria for a long time. At the Cape they said there was no malaria there but at La Paz it was very serious. At La Paz, they said it was only at Loreto. At Loreto they declared that Mulege was full of it. And there it must remain, for we didn't stop at Mulege; so we do not know what the Mulegeños say about it. Later, we picked up the malaria on the other side, ran it down to Topolobambo, and left it there. We would say offhand, never having been to either place, that the malaria is very bad at Mulege and Topolobambo.

A strong, north-pointing peninsula is the outer boundary of Concepción Bay. At the mouth it is three and a quarter miles wide and it extends twenty-two miles southward, varying in width from two to five miles. The eastern shore, along which we collected, is regular in outline, with steep beaches of sand and pebbles and billions of bleaching shells and many clams and great snails. From the shore, the ascent is gradual toward mountains which ridge the little peninsula and protect this small gulf from the Gulf of California. Along the shore are many pools of very salty water, where thousands of fiddler crabs sit by their moist burrows and bubble as one approaches. The beach was beautiful with the pink and white shells of the murex.[1] Sparky found them so beautiful that he collected a washtubful of them and stored them in the hold. And even then, back in Monterey, he found he did not have enough for his friends.

Behind the beach there was a little level land, sandy and dry and covered with cactus and thick brush. And behind that, the rising dry hills. Now again the wild doves were calling among the hills with their song of homesickness. The quality of longing in this sound, the memory response it sets up, is curious and strong. And it has also the quality of a dying day. One wishes to walk toward the sound—to walk on and on toward it, forgetting everything else. Undoubtedly there are sound symbols in the unconscious just as there are visual symbols—sounds that trigger off a response, a little spasm of fear, or a quick lustfulness, or, as with the doves, a nostalgic sadness. Perhaps in our pre-humanity this sound of doves was a signal that the day was over and a night of terror due—a

[1] *Phyllonotus bicolor.*

night which perhaps this time was permanent. Keyed to the visual symbol of the sinking sun and to the odor symbol of the cooling earth, these might all cause the little spasm of sorrow; and with the long response-history, one alone of these symbols might suffice for all three. The smell of a musking goat is not in our experience, but it is in some experience, for smelled faintly, or in perfume, it is not without its effect even on those who have not smelled the passionate gland nor seen the play which follows its discharge. But some great group of shepherd peoples must have known this odor and its result, and must, from the goat's excitement, have taken a very strong suggestion. Even now, a city man is stirred deeply when he smells it in the perfume on a girl's hair. It may be thought that we produce no musk nor anything like it, but this we do not believe. One has the experience again and again of suddenly turning and following with one's eyes some particular girl among many girls, even trotting after her. She may not be beautiful, indeed, often is not. But what are the stimuli if not odors, perhaps above or below the conscious olfactory range? If one follows such an impulse to its conclusion, one is not often wrong. If there be visual symbols, strong and virile in the unconscious, there must be others planted by the other senses. The sensitive places, ball of thumb, ear-lobe, skin just below the ribs, thigh, and lip, must have their memories too. And smell of some spring flowers when the senses thaw, and smell of a ready woman, and smell of reptiles and smell of death, are deep in our unconscious. Sometimes we can say truly, "That man is going to die." Do we smell the disintegrating cells? Do we see the hair losing its luster and uneasy against the scalp, and the skin dropping its tone? We do not know these reactions one by one, but we say, that man or cat or dog or cow is going to die. If the fleas on a dog know it and leave their host in advance, why do not we also know it? Approaching death, the pre-death of the cells, has informed the fleas and us too.

The shallow water along the shore at Concepción Bay was littered with sand dollars, two common species[2] and one[3] very rare. And in the same association, brilliant-red sponge arborescences[4] grew in loose stones in the sand or on the knobs of old coral. These

[2] *Encope californica* and *E. grandis.*

[3] *Clypeaster rotundus.*

[4] *Tedania ignis.*

are the important horizon markers. On other rocks, imbedded in the sand, there were giant hachas, clustered over with tunicates and bearing on their shells the usual small ophiurans and crabs. One of the masked rock-clams had on it a group of solitary corals. Close inshore were many brilliant large snails, the living animals the shells of which had so moved Sparky. In this area we collected from the skiff, leaning over the edge, bringing up animals in a dip-net or spearing them with a small trident, sometimes jumping overboard and diving for a heavier rock with a fine sponge growing on it.

The ice we had taken aboard at La Paz was all gone now. We started our little motor and ran it for hours to cool the ice-chest, but the heat on deck would not permit it to drop the temperature below about thirty-eight degrees F., and the little motor struggled and died often, apparently hating to run in such heat. It sounded tired and sweaty and disgusted. When the evening came, we had fried fish, caught that day, and after dark we lighted the deck and put our reflecting lamp over the side. We netted a serpent-like eel, thinking from its slow, writhing movement through the water that it might be one of the true viperine sea-snakes which are common farther south. Also we captured some flying fish.

We used long-handled dip-nets in the lighted water, and set up the enameled pans so that the small pelagic animals could be dropped directly into them. The groups in the pans grew rapidly. There were *heteronereis* (the free stages of otherwise crawling worms who develop paddle-like tails upon sexual maturity). There were swimming crabs, other free-swimming annelids, and ribbon-fish which could not be seen at all, so perfectly transparent were they. We should not have known they were there, if they had not thrown faint shadows on the bottom of the pans. Placed in alcohol, they lost their transparency and could easily be seen. The pans became crowded with little skittering animals, for each net brought in many species. When the hooded light was put down very near the water, the smallest animals came to it and scurried about in a dizzying dance so rapidly that they seemed to draw crazy lines in the water. Then the small fishes began to dart in and out, snapping up this concentration, and farther out in the shadows the large wise fishes cruised, occasionally swooping and gobbling the small fishes. Several more of the cream-colored spotted snake-eels wriggled near and were netted. They were very snake-like and they had small bright-blue eyes. They did not swim with a beating tail as fishes do, but rather squirmed through the water.

While we worked on the deck, we put down crab-nets on the bottom, baiting them with heads and entrails of the fish we had had for dinner. When we pulled them up they were loaded with large stalk-eyed snails[5] and with sea-urchins having long vicious spines.[6] The colder-water relatives of both these animals are very slow-moving, but these moved quickly and were completely voracious. A net left down five minutes was brought up with at least twenty urchins in it, and all attacking the bait. In addition to the speed with which they move, these urchins are clever and sensitive with their spines. When approached, the long sharp little spears all move and aim their points at the approaching body until the animal is armed like a Macedonian phalanx. The main shafts of the spines were cream-yellowish-white, but a half-inch from the needle-points they were blue-black. The prick of one of the points burned like a bee-sting. They seemed to live in great numbers at four fathoms; we do not know their depth range, but their physical abilities and their voraciousness would indicate a rather wide one. In the same nets we took several dromiaceous crabs, reminiscent of hermits, which had adjusted themselves to life in half the shell of a bivalve, and had changed their body shapes accordingly.

It is probable that no animal tissue ever decays in this water. The furious appetites which abound would make it unlikely that a dead animal, or even a hurt animal, should last more than a few moments. There would be quick death for the quick animal which became slow, for the shelled animal which opened at the wrong time, for the fierce animal which grew timid. It would seem that the penalty for a mistake or an error would be instant death and there would be no second chance.

It would have been good to keep some of the sensitive urchins alive and watch their method of getting about and their method of attack. Indeed, we will never go again without a full-sized observation aquarium into which we can put interesting animals and keep them for some time. The aquaria taken were made with po-

[5] *Strombus* spp.

[6] *Astropyga pulvinata.*

larized glass. Thus, the fish could look out but we could not look in. This, it turned out, was an error on our part.

There are three ways of seeing animals: dead and preserved; in their own habitats for the short time of a low tide; and for long periods in an aquarium. The ideal is all three. It is only after long observation that one comes to know the animal at all. In his natural place one can see the normal life, but in an aquarium it is possible to create abnormal conditions and to note the animal's adaptability or lack of it. As an example of this third method of observation, we can use a few notes made during observation of a small colony of anemones in an aquarium. We had them for a number of months.

In their natural place in the tide pool they are thick and close to the rock. When the tide covers them they extend their beautiful tentacles and with their nettle-cells capture and eat many micro-organisms. When a powerful animal, a small crab for example, touches them, they paralyze it and fold it into the stomach, beginning the digestive process before the animal is dead, and in time ejecting the shell and other indigestible matter. On being touched by an enemy, they fold in upon themselves for protection. We brought a group of these on their own stone into the laboratory and placed them in an aquarium. Cooled and oxygenated sea water was sprayed into the aquarium to keep them alive. Then we gave them various kinds of food, and found that they do not respond to simple touch-stimulus on the tentacles, but have something which is at least a vague parallel to taste-buds, whatever may be the chemical or mechanical method. Thus, protein food was seized by the tentacles, taken and eaten without hesitation; fat was touched gingerly, taken without enthusiasm to the stomach, and immediately rejected; starches were not taken at all—the tentacles touched starchy food and then ignored it. Sugars, if concentrated, seemed actually to burn them so that the tentacles moved away from contact. There did really appear to be a chemical method of differentiation and choice. We circulated the same sea water again and again, only cooling and freshening it. Pure oxygen, introduced into the stomach in bubbles, caused something like drunkenness; the animal relaxed and its reaction to touch was greatly slowed, and sometimes completely stopped for a while. But the reaction to chemical stimulus remained active, although slower. In time, all the microscopic food was removed from the water through constant circulation past the anemones, and then the animals began to change their shapes. Their bodies, which had been thick and fat, grew long and neck-like; from a normal inch in length, they changed to three inches long and very slender. We suspected this was due to starvation. Then one day, after three months, we dropped a small crab into the aquarium. The anemones, moving on their new long necks, bent over and attacked the crab, striking downward like slow snakes. Their normal reaction would have been to close up and draw in their tentacles, but these animals had changed their pattern in hunger, and now we found that when touched on the body, even down near the base, they moved downward, curving on their stalks, while their tentacles hungrily searched for food. There seemed even to be competition among the individuals, a thing we have never seen in a tide pool among anemones. This versatility had never been observed by us and is not mentioned in any of the literature we have seen.

The aquarium is a very valuable extension of shore observation. Quick-eyed, timid animals soon become used to having humans about, and quite soon conduct their business under lights. If we could have put our sensitive urchins in an aquarium, we could have seen how it is that they move so rapidly and how they are stimulated to aim their points at an approaching body. But we preserved them, and of course they lost color and dropped many of their beautiful sharp spines. Also, we could have seen how the great snails are able to consume animal tissue so quickly. As it is, we do not know these things.

19

MARCH 29

Tides had been giving us trouble, for we were now far enough up the Gulf so that the tidal run had to be taken into time consideration. In the evening we had set up a flagged stake at the waterline, so that with glasses we could see from the deck the rise and fall of the tide in relation to the stick. At seven-thirty in the morning the tide was going down from our marker. We had abandoned our tide charts as useless by now, and since we stayed such a short time at each station we could not make new ones. The irregular length of our jumps made it impossible for us to forecast with accuracy from a preceding station. Besides all this, a good, leisurely state of mind had come over us which had nothing to do with the speed and duration of our work. It is very possible to work hard and fast in a leisurely manner, or to work slowly and clumsily with great nervousness.

On this day, the sun glowing on the morning beach made us feel good. It reminded us of Charles Darwin, who arrived late at night on the *Beagle* in the Bay of Valparaiso. In the morning he awakened and looked ashore and he felt so well that he wrote, "When morning came everything appeared delightful. After Tierra del Fuego, the climate felt quite delicious, the atmosphere so dry and the heavens so clear and blue with the sun shining brightly, that all nature seemed sparkling with life."[1] Darwin was not saying how it was with Valparaiso, but rather how it was with him. Being a naturalist, he said, "All nature seemed sparkling with life," but actually it was he who was sparkling. He felt so very fine that he can, in these charged though general adjectives, translate his ecstasy over a hundred years to us. And we can feel how he stretched his muscles in the morning air and perhaps took off his hat—we hope a bowler—and tossed it and caught it.

On this morning, we felt the same way at Concepción Bay.

[1] *Voyage of the Beagle*, Chap. 12, July 23.

"Everything appeared delightful." The tiny waves slid up and down the beach, hardly breaking at all; out in the Bay the pelicans were fishing, flying along and then folding their wings and falling in their clumsy-appearing dives, which nevertheless must be effective, else there would be no more pelicans.

By nine A.M. the water was well down, and by ten seemed to have passed low and to be flowing again. We went ashore and followed the tide down. The beach is steep for a short distance, and then levels out to a gradual slope. We took two species of cake urchins which commingled at one-half to one and one-half feet of water at low tide. The ordinary cake urchin here, with holes, is *Encope californica* Verrill. The grotesquely beautiful keyhole sand dollar[2] was very common here. Finally, there was a rare member of the same group,[3] which we collected unknowingly, and turned out only three individuals of the species when the animals were separated on deck. A little deeper, about two feet submerged, at low tide, a species of cucumber new to us was taken, a flat, sand-encrusted fellow.[4] Giant heart-urchins[5] in some places were available in the thousands. They ranged between two feet and three feet below the surface at low water, and very few were deeper. The greatest number occurred at three feet.

The shore line here is much like that at Puget Sound: in the high littoral is a foreshore of gravel to pebbles to small rocks; in the low littoral, gravelly sand and fine sand with occasional stones below the low tide level. In this zone, with a maximum at four feet, were heavy groves of algae, presumably *Sargassum,* lush and tall, extending to the surface. Except for the lack of eel-grass, it might have been Puget Sound. We took giant stalk-eyed conchs,[6] several species of holothurians and *Cerianthus,* the sand anemone whose head is beautiful but whose encased body is very ugly, like rotting gray cloth. Tiny christened *Cerianthus* "sloppy-guts," and the name stuck. By diving, we took a number of hachas, the huge mussel-like clams. Their shells were encrusted with sponges and tuni-

[2] *Encope grandis* L. Agassiz.

[3] *Clypeaster rotundus* (A. Agassiz).

[4] *Holothuria inhabilis.*

[5] *Meoma grandis.*

[6] *Strombus galeatus.*

cates under which small crabs and snapping shrimps hid themselves. Large scalloped limpets also were attached to the shells of the hachas. This creature closes itself so tightly with its big adductor muscle that a knife cannot penetrate it and the shell will break before the muscle will relax. The best method for opening them is to place them in a bucket of water and, when they open a little, to introduce a sharp, thin-bladed knife and sever the muscle quickly. A finger caught between the closing shells would probably be injured. In many of the hachas we found large, pale, commensal shrimps[7] living in the folds of the body. They are soft-bodied and apparently live there always.

About noon we got under way for San Lucas Cove, and as usual did our preserving and labeling while the boat was moving. Some of the sand dollars we killed in formalin and then set in the sun to dry, and many more we preserved in formaldehyde solution in a small barrel. We had taken a great many of them. Sparky had, by now, filled several sacks with the fine white rose-lined murex shells, explaining, as though he were asked for an explanation, that they would be nice for lining a garden path. In reality, he simply loved them and wanted to have them about.

We passed Mulege, that malaria-ridden town, that town of high port fees—so far as we know—and it looked gay against the mountains, red-roofed and white-walled. We wished we were going ashore there, but the wall of our own resolve kept us out, for we had said, "We will *not* stop at Mulege," and having said it, we could not overcome our own decision. Sparky and Tiny looked longingly at it as we passed; they had come to like the quick excursions into little towns: they found that their Italian was understood for any purposes they had in mind. It was their practice to wander through the streets, carrying their cameras, and in a very short time they had friends. Tony and Tex were foreigners, but Tiny and Sparky were very much at home in the little towns—and they never inquired whose home. This was not reticence, but rather a native tactfulness.

Now that we were engaged in headland navigating, Tiny's and Sparky's work at the wheel had improved, and except when they chased a swordfish (which was fairly often) we were not off course

[7] *Pontonia pinnae.*

more than two or three times during their watch. They had abandoned the compass with relief and blue water was no longer thrust upon them.

At about this time it was discovered that Tex was getting fat, and inasmuch as he was to be married soon after his return, we decided to diet him and put him in a marrying condition. He protested feebly when we cut off his food, and for three days he sneaked food and stole food and cozened us out of food. During the three days of his diet, he probably ate twice as much as he did before, but the idea that he was starving made him so hungry that at the end of the third day he said he couldn't stand it any longer, and he ate a dinner that nearly killed him. Actually, with his thefts of food he had picked up a few pounds during his diet, but always afterwards he shuddered at the memory of those three days. He said, "A man doesn't feel his best when he is starving" and he asked what good it would do him to be married if he were weak and sick.

At five P.M. on March 29 we arrived at San Lucas Cove and anchored outside. The cove, a deep salt-water lagoon, guarded by a large sandspit, has an entrance that might mave been deep enough for us to enter, but the current is strong and there were no previous soundings available. Besides, Tony was nervous about taking his boat into such places. There was another reason for anchoring outside; in the open Gulf where the breeze moves there are no bugs, while if one anchors in still water near the mangroves little visitors come and spend the night. There is one small, beetle-like black fly which crawls down into bed with you and has a liking for very tender places. We had suffered from this fellow when the wind blew over the mangroves to us. This bug hates light, but finds security and happiness under the bedding, nestling over one's kidneys, munching contentedly. His bite leaves a fiery itch; his collective soul is roasting in Hell, if we have any influence in the court of Heaven. After one experience with him, we anchored always a little farther out.

When we came to San Lucas, the tide was flowing and the little channel was a mill-race. It would be necessary to wait for the morning tide. We were eager to see whether on this sandbar, so perfectly situated, we could not find amphioxus, that most primitive of vertebrates. As we dropped anchor a large shark cruised about us, his fin high above the water. We shot at him with a pistol and one shot went through his fin. He cut away like a razor blade and we could hear the hiss of the water. What incredible speed sharks can make when they hurry! We wonder how their greatest speed compares with that of a porpoise. The variations in speed among individuals of these fast-swimming species must be very great too. There must be incredible sharks, like Man o' War or Charlie Paddock, which make other sharks seem slow.

That night we hung the light over the side again and captured some small squid, the usual *heteronereis*, a number of free-swimming crustacea, quantities of crab larvae and the transparent ribbon-fish again. The boys developed a technique for catching flying fish: one jabbed at it with a net, making it fly into the net of another. But even in the nets they were not caught, for they struggled and fluttered away with ease. That night we had a mild celebration of some minor event which did not seem important enough to remember. The pans of animals were still lying on the deck and one of our members, confusing Epsom salts with cracker-crumbs, tried to anesthetize a large pan of holothurians with cracker-meal. The resulting thick paste seemed to have no narcotic qualities whatever.

Late, late in the night we recalled that Horace says fried shrimps and African snails will cure a hangover. Neither was available. And we wonder whether this classical remedy for a time-bridging ailment has been prescribed and tried since classical times. We do not know what snail he refers to, or whether it is a marine snail or an escargot. It is too bad that such imaginative remedies have been abandoned for the banalities of antiacids, heart stimulants, and analgesics. The Bacchic mystery qualified and nullified by a biochemistry which is almost but not quite yet a mystic science. Horace suggests that wine of Cos taken with these shrimps and snails guarantees the remedy. Perhaps it would. In that case, his remedy is in one respect like those unguents used in witchcraft which combine such items as dried babies' brains, frog-eyes, lizards' tongues, and mold from a hanged man's skull with a quantity of good raw opium, and thus serve to stimulate the imagination and the central nervous system at the same time. In our pained discussion at San Lucas Cove we found we had no snails nor shrimps nor wine of Cos. We tore the remedy down to its fundamentals, and decided that it was a good strong dose of proteins and alcohol, so we substituted a new compound—fried fish and a dash of medicinal whisky—and it did the job.

The use of euphemism in national advertising is giving the hangover a bad name. "Over-indulgence" it is called. There is a curious nastiness about over-indulgence. We would not consider over-indulging. The name is unpleasant, and the word "over" indicates that one shouldn't have done it. Our celebration had no such implication. We did *not* drink too much. We drank just enough, and we refuse to profane a good little time of mild inebriety with that slurring phrase "over-indulgence."

There was a reference immediately above to the medicine chest. On leaving Monterey, it may be remembered, we had exhausted the medicine, but no sooner had we put to sea when it was discovered that each one of us, with the health of the whole party in mind, had laid in auxiliary medicine for emergencies. We had indeed, when the good-will of all was assembled, a medicine chest which would not have profaned a fair-sized bar. And the emergencies did occur. Who is to say that an emergency of the soul is not worse than a bad cold? What was good enough for Li-Po was good enough for us. There have been few enough immortals who did not love wine; offhand we cannot think of any and we do not intend to try very hard. The American Indians and the Australian Bushmen are about the only great and intellectual peoples who have not developed an alcoholic liquor and a cult to take care of it. There are, indeed, groups among our own people who have abandoned the use of alcohol, due no doubt to Indian or Bushman blood, but we do not wish to claim affiliation with them. One can imagine such a specimen of Bushman reading this journal and saying, "Why, it was all drinking—beer—and at San Lucas Cove, whisky." So might a night-watchman cry out, "People sleep all the time!" So might a blind man complain, "Among some people there is a pernicious and wicked practice called 'seeing.' This eventually causes death and should be avoided." Actually, with few tribal exceptions, our race has a triumphant alcoholic history and no definite symptoms of degeneracy can be attributed to it. The theory that alcohol is a poison was too easily and too blindly accepted. So it is to some individuals; sugar is poison to others and meat to others. But to the race in general, alcohol has been an anodyne, a warmer of the soul, a strengthener of muscle and spirit. It has given courage to cowards and has made very ugly people attractive. There is a story told of a Swedish tramp, sitting in a ditch on Midsummer Night. He was ragged and dirty and drunk, and he said to himself softly and in wonder, "I am rich and happy and perhaps a little beautiful."

20

MARCH 30

At eight-thirty in the morning the tide was ebbing, uncovering the sand-bar and a great expanse of tidal sand-flat. This flat was made up to a large extent of the broken shells of mollusks. In digging, we found many small clams and a few smooth *Venus*-like clams. We took one very large male fiddler crab. "Sloppy-guts," the *Cerianthus*, was very common here. There were numbers of hermit crabs and many of the swimming crabs with bright-blue claws. These crabs[1] are eaten by Mexicans and are delicious. They swim very rapidly through the water. When we pursued them to the shallows they tried to escape for a time, but soon settled to the bottom and raised their claws to a position not unlike that of a defensive boxer. Their pinch was very painful. When captured and put into a collecting bucket they vented their fury on one another; pinched-off legs and claws littered the buckets on our return. These crabs do not seem to come out of the water as the grapsoids do. Removed from the water, they very soon weaken and lose their fight. Moreover, they do not die as rapidly in fresh water as do most other crabs. Perhaps, living in the lagoons which sometimes must be almost brackish, they have achieved a tolerance for fresh water greater than that of other crabs; greater indeed, although it is not much of a trick, than that of a certain biologist who shall be nameless.

This varying threshold of tolerance is always an astonishing thing.

Amphioxus ordinarily lives on the seaward side of a sand-bar and in sub-tidal water; or, at least, in sand bared by only the lowest of tides. We dug for them here and took only a few weak ones. It was not a very low tide and these were very possibly stragglers. It is probable that an extremely low tide would expose a level in which a great many of them live. The capture of these animals is

[1] *Callinectes bellicosus.*

exciting and requires speed. They are perfectly streamlined and partly transparent. Also, they are extremely nervous. Sometimes if the sand is struck with a shovel they will jump out and then frantically wriggle to get under the sand again—which they readily do. They are able to move through sand and even under it with great rapidity. We turned over the sand and leaped at them before they could escape. There used to be very many of them at Balboa Beach in Southern California, but channel dredging and perhaps the great number of motor boats have made them rare. And they are very interesting animals, being almost the dividing point between vertebrates and invertebrates. Usually one to three inches long and shuttle-shaped, they are perfectly built to slip through the sand without resistance.

The bar was rich with clams, many small *Chione*, and some small razor clams. We extracted the *Cerianthus* from their sloppy casings and found a great many tiny commensal sipunculid worms[2] in the smooth inner linings of the cases. These were able to extend themselves so far that they seemed like hairs, or to retract until they were like tiny peanuts. We did not find commensal pea crabs in the linings, as we had thought we would.

San Lucas Cove is nearly slough-like. The water gets very warm and probably very stale. It is exposed to a deadly sun and is so shallow that the water is soupy. This very quality of probable high salinity and warmth made it very difficult to preserve the *Cerianthus* in an expanded state. The small bunodids are easily anesthetized in Epsom salts, but *Cerianthus*, after six to eight hours of concentrated Epsom-salts solution, and even standing in pans under the hot sun, were able to retract rapidly and violently by expelling water from the aboral pore when the preserving liquid touched them. Sooner or later we will find the perfect method for anesthetizing anemones, but it has not yet been found. There is hope that cold may work as the anesthetic, if we can force absorption of formalin while the animals are relaxed with dry ice. But a great deal of experimenting is necessary, for if too cold they do not receive the formalin, and if too warm they retract on contact with it.

Back on board at about eleven-thirty we sailed for San Carlos Bay. We did not plan to stop at Santa Rosalia. It is a fairly large

[2] *Phascolosoma hesperum.*

town which has long been supported by copper mines in the neighborhood, under the control of a French company. A little feeling of hurry was creeping upon us, for by now we had begun to see the magnitude of the job we had undertaken, and to realize that with the limited time and the more than limited equipment and personnel, we could not make much of a job of it. Our time was going fast. Much as Sparky and Tiny wished to continue their research and shore collecting at Santa Rosalia, we sailed on past it. And it looked, from the sea at least, to be less Mexican than other towns. Perhaps that was because we knew it was run by a French company. A Mexican town grows out of the ground. You cannot conceive its never having been there. But Santa Rosalia looked "built." There were industrial works of large size visible, loading trestles, and piles of broken rock. The mountains rose behind the town, burned almost white, and the green about the houses and the red roofs were in startling contrast. Sparky had the wheel as we went by, and his left hand was heavy. It required a definite effort of will for him to keep the course off shore.

At about six P.M. we came to San Carlos Bay, a curious landlocked curve with an inner shallow bay. There is good anchorage for small craft in the outer bay, with five to seven fathoms of water. The inner bay, or lagoon, has a sand beach on all sides. We intended to collect on the heavy boulders on the inner, or eastern, shore. There might be, we thought, a contrasting fauna to that of the tide flats of morning. This beach was piled high with rotting seaweed, left by some fairly recent storm perhaps. Or possibly this beach is at the end of some current-cycle, so that a high tide deposits great amounts of torn weed. There is such a beach at San Antonio del Mar on the western shore of Lower California, about sixty miles south of Ensenada. The debris from ships from hundreds of miles around is piled on this beach—mountains of sea-washed boxes and crates, logs and lumber, great whitened piles of it, mixed in with bottles and cans and pieces of clothing. It is the termination of some great sweeping in the Pacific.

Here at San Carlos there was little human debris; so very few boats pass up the Gulf this far and the people so prize planed wood and cans that such things would be picked up very quickly. In the decaying weed were myriads of flies and beachhoppers working on this endless food supply. But in spite of their incredible numbers, we were able to catch only a few of the hoppers; they were too fast for us. Again we felt that here in the Gulf a little extra is added to the protection of animals. They are extra-fast, they are extra-armored, they seem to sting and pinch and bite worse than animals in other places. In the sand we found some clams rather like the Pismo clams of California, but shiny brown to black; also some ribbed mussel-like clams.[3] On the rocks we took two species of chitons and some new snails and crabs. There were blue, sharp-spined urchins and a number of flatworms. The flatworms are hard to catch, for they flow over the rocks like quicksilver. Also they are impossible to preserve well; many of them simply dissolve in the preservative, while others roll up tightly. *Heliaster*, the sun-star, was here, but he had continued to shrink and was quite small this far up in the Gulf. Under the sand there were a great number of heart-urchins.

That night, using the shaded lamp hung over the side, we had a great run of transparent fish, including a type we had not seen before. We took another squid, a larval mantis-shrimp, and the usual *heteronereis* and crustacea.

[3] *Carditamera affinis.*

21

MARCH 31

The tide was very poor this morning, only two and a half to three feet below the uppermost line of barnacles. We started about ten o'clock and had a little collecting under water, but soon the wind got up and so ruffled the surface that we could not see what we were doing. To a certain extent this was a good thing. Not being able to get into the low littoral, where no doubt the spectacular spiny lobsters would have distracted us, we were able to make a more detailed survey of the upper region. One fact increasingly emerged: the sulphury-green and black cucumber[1] is the most ubiquitous shore animal of the Gulf of California, with *Heliaster*, the sun-star, a close second. These two are found nearly everywhere. In this region at San Carlos, Sally Lightfoot lives highest above the ordinary high tide, together with a few *Ligyda occidentalis*, a cockroach-like crustacean. Attached to the rocks and cliffsides, high up and fully exposed to this deadly sun, were barnacles and limpets, so placed that they must experience only occasional immersion, although they may be often dampened by spray. Under rocks and boulders, in the next association lower down, were the mussel-like ruffled clams and the brown chitons, many cucumbers, a few *Heliasters,* and only two species of brittle-stars—another common species, *Ophiothrix spiculata,* we did not find here although we had seen it everywhere else. In this zone verrucose anemones were growing under overhangs on the sides of rocks and in pits in the rocks. There were also a few starfish[2]; garbanzo clams were attached to the rock undersides by the thousands together with club urchins. Farther down in a new zone was a profusion of sponges of a number of species, including a beautiful blue sponge. There were octopi[3] here, and one species of chiton; there were many large purple ur-

[1] *Holothuria lubrica.*

[2] *Astrometis sertulifera.*

[3] *Octopus bimaculatus.*

chins, although no specimens were taken, and heart-urchins in the sand and between the rocks. There were some sipunculids and a great many tunicates.

We found extremely large sponges, a yellow form (probably *Cliona*) superficially resembling the Monterey *Lissodendoryx noxiosa,* and a white one, *Steletta,* of the wicked spines. There were brilliant-orange nudibranchs, giant terebellid worms, some shell-less air-breathing (pulmonate) snails, a ribbon-worm, and a number of solitary corals. These were the common animals and the ones in which we were most interested, for while we took rarities when we came upon them in normal observation, our interest lay in the large groups and their associations—the word "association" implying a biological assemblage, all the animals in a given habitat.

It would seem that the commensal idea is a very elastic thing and can be extended to include more than host and guest; that certain kinds of animals are often found together for a number of reasons. One, because they do not eat one another; two, because these different species thrive best under identical conditions of wave-shock and bottom; three, because they take the same kinds of food, or different aspects of the same kinds of food; four, because in some cases the armor or weapons of some are protection to the others (for instance, the sharp spines of an urchin may protect a tide-pool johnny from a larger preying fish); five, because some actual commensal partition of activities may truly occur. Thus the commensal tie may be loose or very tight and some associations may partake of a real thigmotropism.

Indeed, as one watches the little animals, definite words describing them are likely to grow hazy and less definite, and as species merges into species, the whole idea of definite independent species begins to waver, and a scale-like concept of animal variations comes to take its place. The whole taxonomic method in biology is clumsy and unwieldy, shot through with the jokes of naturalists and the egos of men who wished to have animals named after them.

Originally the descriptive method of naming was not so bad, for every observer knew Latin and Greek well and was able to make out the descriptions. Such knowledge is fairly rare now and not even requisite. How much easier if the animals bore numbers to which the names were auxiliary! Then, one knowing that the phylum Arthropoda was represented by the roman figure *VI,* the class

Crustacea by a capital *B*, order by arabic figure *13*, and genus and species by a combination of small letters, would with little training be able to place the animals in his mind much more quickly and surely than he can now with the descriptive method tugged bodily from a discarded antiquity.

As we ascended the Gulf it became more sparsely inhabited; there were fewer of the little heat-struck *rancherias*, fewer canoes of fishing Indians. Above Santa Rosalia very few trading boats travel. One would be really cut off up here. And yet here and there on the beaches we found evidences of large parties of fishermen. On one beach there were fifteen or twenty large sea-turtle shells and the charcoal of a bonfire where the meat had been cooked or smoked. In this same place we found also a small iron harpoon which had been lost, probably the most valued possession of the man who had lost it. These Indians do not seem to have firearms; probably the cost of them is beyond even crazy dreaming. We have heard that in some of the houses are the treasured weapons of other times, muskets, flintlocks, old long muzzle-loaders kept from generation to generation. And one man told us of finding a piece of Spanish armor, a breastplate, in an Indian house.

There is little change here in the Gulf. We think it would be very difficult to astonish these people. A tank or a horseman armed cap-a-pie would elicit the same response—a mild and dwindling interest. Food is hard to get, and a man lives inward, closely related to time; a cousin of the sun, at feud with storm and sickness. Our products, the mechanical toys which take up so much of our time, preoccupy and astonish us so, would be considered what they are, rather clever toys but not related to very real things. It would be interesting to try to explain to one of these Indians our tremendous projects, our great drives, the fantastic production of goods that can't be sold, the clutter of possessions which enslave whole populations with debt, the worry and neuroses that go into the rearing and educating of neurotic children who find no place for themselves in this complicated world; the defense of the country against a frantic nation of conquerors, and the necessity for becoming frantic to do it; the spoilage and wastage and death necessary for the retention of the crazy thing; the science which labors to acquire knowledge, and the movement of people and goods contrary to the knowledge obtained. How could one make an Indian understand the medicine which labors to save a syphilitic, and the gas and bomb to kill him when he is well, the armies which build health so that death will be more active and violent. It is quite possible that to an ignorant Indian these might not be evidences of a great civilization, but rather of inconceivable nonsense.

It is not implied that this fishing Indian lives a perfect or even a very good life. A toothache may be to him a terrible thing, and a stomachache may kill him. Often he is hungry, but he does not kill himself over things which do not closely concern him.

A number of times we were asked, Why do you do this thing, this picking up and pickling of little animals? To our own people we could have said any one of a number of meaningless things, which by sanction have been accepted as meaningful. We could have said, "We wish to fill in certain gaps in the knowledge of the Gulf fauna." That would have satisfied our people, for knowledge is a sacred thing, not to be questioned or even inspected. But the Indian might say, "What good is this knowledge? Since you make a duty of it, what is its purpose?" We could have told our people the usual thing about the advancement of science, and again we would not have been questioned further. But the Indian might ask, "Is it advancing, and toward what? Or is it merely becoming complicated? You save the lives of children for a world that does not love them. It is our practice," the Indian might say, "to build a house before we move into it. We would not want a child to escape pneumonia, only to be hurt all its life." The lies we tell about our duty and our purposes, the meaningless words of science and philosophy, are walls that topple before a bewildered little "why." Finally, we learned to know why we did these things. The animals were very beautiful. Here was life from which we borrowed life and excitement. In other words, we did these things because it was pleasant to do them.

We do not wish to intimate in any way that this hypothetical Indian is a noble savage who lives in logic. His magics and his techniques and his teleologies are just as full of nonsense as ours. But when two people, coming from different social, racial, intellectual patterns, meet and wish to communicate, they must do so on a logical basis. Clavigero discusses what seems to our people a filthy practice of some of the Lower California Indians. They were always hungry, always partly starved. When they had meat, which was a rare thing, they tied pieces of string to each mouthful, then ate it, pulled it up and ate it again and again, often passing it from hand to hand. Clavigero found this a disgusting practice. It is rather like the Chinese being ridiculed for eating twenty-year-old eggs who said, "Your cheese is rotten milk. You like rotten milk—we like rotten eggs. We are both silly."

Costume on the *Western Flyer* had degenerated completely. Shirts were no longer worn, but the big straw hats were necessary. On board we went barefoot, clad only in hats and trunks. It was easy then to jump over the side to freshen up. Our clothes never got dry; the salt deposited in the fibers made them hygroscopic, always drawing the humidity. We washed the dishes in hot salt water, so that little crystals stuck to the plates. It seemed to us that the little salt adhering to the coffee pot made the coffee delicious. We ate fish nearly every day: bonito, dolphin, sierra, red snappers. We made thousands of big fat biscuits, hot and unhealthful. Twice a week Sparky created his magnificent spaghetti. Unbelievable amounts of coffee were consumed. One of our party made some lemon pies, but the quarreling grew bitter over them; the thievery, the suspicion of favoritism, the vulgar traits of selfishness and perfidy those pies brought out saddened all of us. And when one of us who, from being the most learned should have been the most self-controlled, took to hiding pie in his bed and munching it secretly when the lights were out, we decided there must be no more lemon pie. Character was crumbling, and the law of the fang was too close to us.

One thing had impressed us deeply on this little voyage: the great world dropped away very quickly. We lost the fear and fierceness and contagion of war and economic uncertainty. The matters of great importance we had left were not important. There must be an infective quality in these things. We had lost the virus, or it had been eaten by the anti-bodies of quiet. Our pace had slowed greatly; the hundred thousand small reactions of our daily world were reduced to very few. When the boat was moving we sat by the hour watching the pale, burned mountains slip by. A playful swordfish, jumping and spinning, absorbed us completely. There was time to observe the tremendous minutiae of the sea. When a school of fish went by, the gulls followed closely. Then the water was littered with feathers and the scum of oil. These fish were much too large for the gulls to kill and eat, but there is much more to a school of fish than the fish themselves. There is constant vomiting; there are the hurt and weak and old to cut out; the smaller prey on which the school feeds sometimes escape and die; a moving school is like a moving camp, and it leaves a camp-like debris behind it on which the gulls feed. The sloughing skins coat the surface of the water with oil.

At six P.M. we made anchorage at San Francisquito Bay. This cove-like bay is about one mile wide and points to the north. In the southern part of the bay there is a pretty little cove with a narrow entrance between two rocky points. A beach of white sand edges this cove, and on the edge of the beach there was a poor Indian house, and in front of it a blue canoe. No one came out of the house. Perhaps the inhabitants were away or sick or dead. We did not go near; indeed, we had a strong feeling of intruding, a feeling sharp enough even to prevent us from collecting on that little inner bay. The country hereabouts was stony and barren, and even the brush had thinned out. We anchored in four fathoms of water on the westerly side of the bay, then went ashore immediately and set up our tide stake at the water's edge, with a bandanna on it so we could see it from the boat. The wind was blowing and the water was painfully cold. The tide had dropped two feet below the highest line of barnacles. Three types of crabs[4] were common here. There were many barnacles and great limpets and two species of snails, *Tegula* and a small *Purpura*. There were many large smooth brown chitons, and a few bristle-chitons. Farther down under the rocks were great anastomosing masses of a tube-worm with rusty red gills,[5] some tunicates, *Astrometis,* and the usual holothurians.

Tiny found the shell of a fine big lobster,[6] newly cleaned by isopods. The isopods and amphipods in their millions do a beautiful job. It is common to let them clean skeletons designed for study.

[4] *Pachygrapsus crassipes, Geograpsus lividus,* and, under the rocks, *Petrolisthes nigrunguiculatus,* a porcelain crab.

[5] *Salmacina.*

[6] Apparently the northern *Panulirus interruptus.*

A dead fish is placed in a jar having a cap pierced with holes just large enough to permit the entrance of the isopods. This is lowered to the bottom of a tide pool, and in a very short time the skeleton is clean of every particle of flesh, and yet is articulated and perfect.

The wind blew so and the water was so cold and ruffled that we did not stay ashore for very long. On board, we put down the baited bottom nets as usual to see what manner of creatures were crawling about there. When we pulled up one of the nets, it seemed to be very heavy. Hanging to the bottom of it on the outside was a large horned shark.[7] He was not caught, but had gripped the bait through the net with a bulldog hold and he would not let go. We lifted him unstruggling out of the water and up onto the deck, and still he would not let go. This was at about eight o'clock in the evening. Wishing to preserve him, we did not kill him, thinking he would die quickly. His eyes were barred, rather like goat's eyes. He did not struggle at all, but lay quietly on the deck, seeming to look at us with a baleful, hating eye. The horn, by the dorsal fin, was clean and white. At long intervals his gill-slits opened and closed but he did not move. He lay there all night, not moving, only opening his gill-slits at great intervals. The next morning he was still alive, but all over his body spots of blood had appeared. By this time Sparky and Tiny were horrified by him. Fish out of water should die, and he didn't die. His eyes were wide and for some reason had not dried out, and he seemed to regard us with hatred. And still at intervals his gill-slits opened and closed. His sluggish tenacity had begun to affect all of us by this time. He was a baleful personality on the boat, a sluggish, gray length of hatred, and the blood spots on him did not make him more pleasant. At noon we put him into the formaldehyde tank, and only then did he struggle for a moment before he died. He had been out of the water for sixteen or seventeen hours, had never fought or flopped a bit. The fast and delicate fishes like the tunas and mackerels waste their lives out in a complete and sudden flurry and die quickly. But about this shark there was a frightful quality of stolid, sluggish endurance. He had come aboard because he had grimly fastened on the bait and would not release it, and he lived because he would

[7] *Gyropleurodus* of the Heterodontidae.

not release life. In some earlier time he might have been the basis for one of those horrible myths which abound in the spoken literature of the sea. He had a definite and terrible personality which bothered all of us, and, as with the sea-turtle, Tiny was shocked and sick that he did not die. This fish, and all the family of the Heterodontidae, ordinarily live in shallow, warm lagoons, and, although we do not know it, the thought occurred to us that sometimes, perhaps fairly often, these fish may be left stranded by a receding tide so that they may have developed the ability to live through until the flowing tide comes back. The very sluggishness in that case would be a conservation of vital energy, whereas the beautiful and fragile tuna make one frantic rush to escape, conserving nothing and dying immediately.

Within our own species we have great variation between these two reactions. One man may beat his life away in furious assault on the barrier, where another simply waits for the tide to pick him up. Such variation is also observable among the higher vertebrates, particularly among domestic animals. It would be strange if it were not also true of the lower vertebrates, among the individualistic ones anyway. A fish, like the tuna or the sardine, which lives in a school, would be less likely to vary than this lonely horned shark, for the school would impose a discipline of speed and uniformity, and those individuals which would not or could not meet the school's requirements would be killed or lost or left behind. The overfast would be eliminated by the school as readily as the overslow, until a standard somewhere between the fast and slow had been attained. Not intending a pun, we might note that our schools have to some extent the same tendency. A Harvard man, a Yale man, a Stanford man—that is, the ideal—is as easily recognized as a tuna, and he has, by a process of elimination, survived the tests against idiocy and brilliance. Even in physical matters the standard is maintained until it is impossible, from speech, clothing, haircuts, posture, or state of mind, to tell one of these units of his school from another. In this connection it would be interesting to know whether the general collectivization of human society might not have the same effect. Factory mass production, for example, requires that every man conform to the tempo of the whole. The slow must be speeded up or eliminated, the fast slowed down. In a thoroughly collectivized state, mediocre efficiency might be very great, but only through the complete elimination of the swift, the clever, and the intelligent, as well as the incompetent. Truly collective man might in fact abandon his versatility. Among school animals there is little defense technique except headlong flight. Such species depend for survival chiefly on tremendous reproduction. The great loss of eggs and young to predators is the safety of the school, for it depends for its existence on the law of probability that out of a great many which start some will finish.

It is interesting and probably not at all important to note that when a human state is attempting collectivization, one of the first steps is a frantic call by the leaders for an increased birth rate—replacement parts in a shoddy and mediocre machine.

Our interest had been from the first in the common animals and their associations, and we had not looked for rarities. But it was becoming apparent that we were taking a number of new and unknown species. Actually, more than fifty species undescribed at the time of capture will have been taken. These will later have been examined, classified, described, and named by specialists. Some of them may not be determined for years, for it is one of the little by-products of the war that scientific men are cut off from one another. A Danish specialist in one field is unable to correspond with his colleague in California. Thus some of these new animals may not be named for a long time. We have listed in the Appendix those already specified and indicated in so far as possible those which have not been worked on by specialists.

Dr. Rolph Bolin, ichthyologist at the Hopkins Marine Station, found in our collection what we thought to be a new species of commensal fish which lives in the anus of a cucumber, flipping in and out, possibly feeding on the feces of the host but more likely merely hiding in the anus from possible enemies. This fish later turned out to be an already named species, but, carrying on the ancient and disreputable tradition of biologists, we had hoped to call it by the euphemistic name *Proctophilus winchellii.*

There are some marine biologists whose chief interest is in the rarity, the seldom seen and unnamed animal. These are often wealthy amateurs, some of whom have been suspected of wishing to tack their names on unsuspecting and unresponsive invertebrates. The passion for immortality at the expense of a little beast must be very great. Such collectors should to a certain extent be regarded as in the same class with those philatelists who achieve a great emotional stimulation from an unusual number of perforations or a misprinted stamp. The rare animal may be of individual interest, but he is unlikely to be of much consequence in any ecological picture. The common, known, multitudinous animals, the red pelagic lobsters which litter the sea, the hermit crabs in their billions, scavengers of the tide pools, would by their removal affect the entire region in widening circles. The disappearance of plankton, although the components are microscopic, would probably in a short time eliminate every living thing in the sea and change the whole of man's life, if it did not through a seismic disturbance of balance eliminate all life on the globe. For these little animals, in their incalculable numbers, are probably the base food supply of the world. But the extinction of one of the rare animals, so avidly sought and caught and named, would probably go unnoticed in the cellular world.

Our own interest lay in relationships of animal to animal. If one observes in this relational sense, it seems apparent that species are only commas in a sentence, that each species is at once the point and the base of a pyramid, that all life is relational to the point where an Einsteinian relativity seems to emerge. And then not only the meaning but the feeling about species grows misty. One merges into another, groups melt into ecological groups until the time when what we know as life meets and enters what we think of as non-life: barnacle and rock, rock and earth, earth and tree, tree and rain and air. And the units nestle into the whole and are inseparable from it. Then one can come back to the microscope and the tide pool and the aquarium. But the little animals are found to be changed, no longer set apart and alone. And it is a strange thing that most of the feeling we call religious, most of the mystical outcrying which is one of the most prized and used and desired reactions of our species, is really the understanding and the attempt to say that man is related to the whole thing, related inextricably to all reality, known and unknowable. This is a simple thing to say, but the profound feeling of it made a Jesus, a St. Augustine, a St. Francis, a Roger Bacon, a Charles Darwin, and an Einstein. Each of them in his own tempo and with his own voice discovered and reaffirmed with astonishment the knowledge that all things are one thing and that one thing is all things—plankton, a shimmering phosphorescence on the sea and the spinning planets and an expanding universe, all bound together by the elastic string of time. It is advisable to look from the tide pool to the stars and then back to the tide pool again.

22

APRIL 1

Without the log we should have lost track of the days of the week, were it not for the fact that Sparky made spaghetti on Thursdays and Sundays. We think he did this by instinct, that he could come out of a profound amnesia, and if he felt an impulse to make spaghetti, it would be found to be either Thursday or Sunday. On Monday we sailed for Angeles Bay, which was to be our last station on the Peninsula. The tides were becoming tremendous, and while the tidal bore of the Colorado River mouth was still a long way off, Tony was already growing nervous about it. During the trip between San Francisquito and Angeles Bay, we worried again over the fact that we were not taking photographs. As has been said, no one was willing to keep his hands dry long enough to use the cameras. Besides, none of us knew much about cameras. But it was a constant source of bad conscience to us.

On this day it bothered us so much that we got out the big camera and began working out its operation. We figured everything except how to put the shutter curtain back to a larger aperture without making an exposure. Several ways were suggested and, as is often the case when more than one method is possible, an argument broke out which left shutters and cameras behind. This was a good one. Everyone except Sparky and Tiny, who had the wheel, gathered on the hatch around the camera, and the argument was too much for the steersmen. They sent down respectful word that either we should bring the camera up where they could hear the argument, or they would abandon their posts. We suggested that this would be mutiny. Then Sparky explained that on an Italian fishing boat in Monterey mutiny, far from being uncommon, was the predominant state of affairs, and that he and Tiny would rather mutiny than not. We took the camera up on the deckhouse and promptly forgot it in another argument.

Except for a completely worthless lot of 8-mm. movie film, this was the closest we came to taking pictures. But some day we shall succeed.

Angeles Bay is very large—twenty-five square miles, the *Coast Pilot* says. It is land-locked by fifteen islands, between several of which there is entrance depth. This is one of the few harbors in the whole Gulf about which the *Coast Pilot* is willing to go out on a limb. The anchorage in the western part of the bay, it says, is safe from all winds. We entered through a deep channel between Red Point and two small islets, pulled into eight fathoms of water near the shore, and dropped our anchor. The *Coast Pilot* had not mentioned any settlement, but here there were new buildings, screened and modern, and on a tiny airfield a plane sat. It was an odd feeling, for we had been a long time without seeing anything modern. Our feeling was more of resentment than of pleasure. We went ashore about three-thirty in the afternoon, and were immediately surrounded by Mexicans who seemed curious and excited about our being there. They were joined by three Americans who said they had flown in for the fishing, and they too seemed very much interested in what we wanted until they were convinced it was marine animals. Then they and the Mexicans left us severely alone. Perhaps we had been hearing too many rumors: it was said that many guns were being run over the border for the trouble that was generally expected during the election. The fishermen did not look like fishermen, and Mexicans and Americans were *too* interested in us until they discovered what we were doing and too uninterested after they had found out. Perhaps we imagined it, but we had a strong feeling of secrecy about the place. Maybe there really were gold mines there and new buildings for recent development. A road went northward from there to San Felipe Bay, we were told. The country was completely parched and desolate, but half-way up a hill we could see a green spot where a spring emerged from a mountain. It takes no more than this to create a settlement in Lower California.

We went first to collect on a bouldery beach on the western side of the bay, and found it fairly rich in fauna. The highest rocks were peopled by anemones, cucumbers, sea-cockroaches and some small porcellanids. There were no Sally Lightfoots visible, in fact no large crabs at all, and only a very few small members of *Heliaster*. The dominant animal here was a soft marine pulmonate which occurred in millions on and under the rocks. We took several hundred of them. There were some chitons, both the smooth brown *Chiton virgulatus* and the fuzzy *Acanthochitona exquisitus*. We saw fine big clusters of the minute tube-worm, *Salmacina*, and there were a great many flatworms oozing along on the undersides of the rocks like drops of spilled brown sirup. Under the rocks we found two octopi, both *Octopus bimaculatus*. They are very clever and active in escaping, and when finally captured they grip the hand and arm with their little suckers, and, if left for any length of time, will cause small blood blisters or, rather, what in another field are called "monkey bites." Under the water and apparently below the ordinary tidal range were brilliant-yellow *Geodia* and many examples of another sponge of magnificent shape and size and color. This last (erect colonies of the cosmopolitan *Cliona celata*, more familiar as a boring sponge) was a reddish pink and stood high and vase-like, some of them several feet in diameter. Most of them were perfectly regular in shape. We took a number of them, dried some out of formaldehyde, and preserved others. The algal zonation on this slope was sharp and apparent—a *Sargassum* was submerged two or three feet at ebb. The rocks in the intertidal were perfectly smooth and bare but below this *Sargassum johnstonii*, in deeper water, there was a great zone of flat, frond-like alga, *Padina durvillaei*.[1] The wind rippled the surface badly but when an occasional lull came we could look down into this deeper water. It did not seem rich in life except for the algae, but then we were unable to turn over the rocks on the bottom.

While we collected, our fishermen rowed aimlessly about, and in our suspicious state of mind they seemed to be more anxious to appear to be fishing than actually to be fishing. We have little doubt that we were entirely wrong about this, but the place breathed suspicion, and no other place had been like that.

We went back on board and deposited our catch, then took the skiff to the sand flats on the northern side of the bay. It was a hard, compact mud sand with a long shallow beach, and it was heavy and difficult to dig into. We took there a number of *Chione* and *Tivela* clams and one poor half-dead amphioxus. Again the tide was not low enough to reach the real habitat of amphioxus, but if there was one stray in the high area, there must be many to be taken on an extremely low tide. We found a number of long tur-

[1] Determinations by Dr. E. Yale Dawson of the Department of Botany, University of California.

reted snails carrying commensal anemones on their shells. On this flat there were a number of imbedded small rocks, and these were rich with animals. There were rock-oysters on them and large highly ornamented limpets and many small snails. Tube-worms clustered on these rocks with pea crabs commensal in the tubes. One fair-sized octopus (not *bimaculatus*) had his home under one of these rocks. These small stones must have been havens in the shifting sands for many animals. The fine mud-like sand would make locomotion difficult except for specially equipped animals, and the others clustered to the rocks where there was footing and security.

The tide began to flow rapidly and the winds came up and we went back to the *Western Flyer*. When we were on board we saw a ship entering the harbor, a big green sailing schooner with her sails furled, coming in under power. She did not approach us, but came to anchorage about as far from us as she could. She was one of those incredible Mexican Gulf craft; it is impossible to say how they float at all and, once floating, how they navigate. The seams are sprung, the paint blistered away, ironwork rusted to lace, decks warped and sagging, and, it is said, so dirty and bebugged that if the cockroaches were not fed, or were in any way frustrated or insulted, they would mutiny and take the ship—and, as one Mexican sailor said, "probably sail her better than the master."

Once the anchor of this schooner was down there was no further sign of life on her and there was no sign of life from the buildings ashore either. The little plane sat in its runway and the houses seemed vacant. We had been asked how long we would remain and had said, until the next morning. Now we felt, curiously enough, that we were interfering with something, that some kind of activity would start only when we left. Again, we were probably all wrong, but it is strange that every one of us caught a sinister feeling from the place. Unless the wind was up, or the anchorage treacherous, we ordinarily kept no anchor watch, but this night the boys got up a number of times and were restless. As with the werewolf, we were probably believing our own imaginations. For a short time in the evening there were lights ashore and then they went out. The schooner did not even put up a riding light, but lay completely dark on the water.

23

APRIL 2

We started early and moved out through the channel to the Gulf. It was not long before we could make out Sail Rock far ahead, with Guardian Angel Island to the east of it. Sail Rock looks exactly like a tall Marconi sail in the distance. It is a high, slender pyramid, so whitened with guano that it catches the light and can be seen for a great distance. Because of its extreme visibility it must have been a sailing point for many mariners. It is more than 160 feet high, rises to a sharp point, and there is deep water close in to it. With lots of time, we would have collected at its base, but we were aimed at Puerto Refugio, at the upper end of Guardian Angel Island. We did take some of our usual moving pictures of Sail Rock, and they were even a little worse than usual, for there was laundry drying on a string and the camera was set up behind it. When developed, the film showed only an occasional glimpse of Sail Rock, but a very lively set of scenes of a pair of Tiny's blue and white shorts snapping in the breeze. It is impossible to say how bad our moving pictures were—one film laboratory has been eager to have a copy of the film, for it embodies in a few thousand feet, so they say, every single thing one should not do with a camera. As an object lesson to beginners they think it would be valuable. If we took close-ups of animals, someone was in the light; the aperture was always too wide or too narrow; we made little jerky pan shots back and forth; we have one of the finest sequences of unadorned sky pictures in existence—but when there was something to take about which we didn't care, we got it perfectly. We dare say there is not in the world a more spirited and beautiful picture of a pair of blue and white shorts than that which we took passing Sail Rock.

The long, snake-like coast of Guardian Angel lay to the east of us; a desolate and fascinating coast. It is forty-two miles long, ten miles wide in some places, waterless and uninhabited. It is said to be crawling with rattlesnakes and iguanas, and a persistent rumor of gold comes from it. Few people have explored it or even gone more than a few steps from the shore, but its fine harbor, Puerto Refugio, indicates by its name that many ships have clung to it in storms and have found safety there. Clavigero calls the island both "Angel de la Guardia" and "Angel Custodio," and we like this latter name better.

The difficulties of exploration of the island might be very great, but there is a drawing power about its very forbidding aspect—a Golden Fleece, and the inevitable dragon, in this case rattlesnakes, to guard it. The mountains which are the backbone of the island rise to more than four thousand feet in some places, sullen and desolate at the tops but with heavy brush on the skirts. Approaching the northern tip we encountered a deep swell and a fresh breeze. The tides are very large here, fourteen feet during our stay, and that not an extreme tide at all. It is probable that a seventeen-foot tide would not be unusual here. Puerto Refugio is really two harbors connected by a narrow channel. It is a safe, deep anchorage, the only danger lying in the strength and speed of the tidal current, which puts a strain on the anchor tackle. It was so strong, indeed, that we were not able to get weighted nets to the bottom; they pulled out sideways in the water and sieved the current of weed and small animals, so that catch was fairly worthwhile anyway.

We took our time getting firm anchorage, and at about three-thirty P.M. rowed ashore toward a sand and rubble beach on the southeastern part of the bay. Here the beach was piled with debris: the huge vertebrae of whales scattered about and piles of broken weed and skeletons of fishes and birds. On top of some low bushes which edged the beach there were great nests three to four feet in diameter, pelican nests perhaps, for there were pieces of fish bone in them, but all the nests were deserted—whether they were old or it was out of season we do not know. We are so used to finding on the beaches evidence of man that it is strange and lonely and frightening to find no single thing that man has touched or used. Tiny and Sparky made a small excursion inland, not over several hundred yards from the shore, and they came back subdued and quiet. They had not seen any rattlesnakes, nor did they want to. The beach was alive with hoppers feeding on the refuse, but the coarse sand was not productive of other animal life. The tide was falling, and we walked around a rocky point to the westward and came into a bouldery flat where the collecting was very rich. The receding water had left many small tide pools. The smoothness of the rocks indicated a fairly strong surf; they were dangerously slippery, and Sally Lightfoots and *Pachygrapsus* both scuttled about. As we moved out toward the entrance of the harbor, the boulders became larger and smoother, and then there was a sudden change to unbroken reef, and the smooth rocks gave way to barnacle- and weed-covered stones. The tide was down about ten feet now, exposing the lower tide pools, rich and beautiful with sponges and corals and small pleasant algae. We tried to cover as much territory as possible, but again and again found ourselves fascinated by some small and perfect pool, like a set stage, peopled with broken-back shrimps and small masked crabs.

The point itself was jagged volcanic rock in which there were high mysterious caves. Entering one, we noticed a familiar smell, and a moment later recognized it. For the sound of our voices alarmed myriads of bats, and their millions of squeaks sounded like rushing water. We threw stones in to try to dislodge some, but they would not brave the daylight, and only squeaked more fiercely.

As evening approached, it grew quite cold. Our hands were torn from the long collecting day and we were glad when it was too dark to work any more. We had taken great numbers of animals. There was an echiuroid worm with a spoon-shaped proboscis, found loose under the rocks; many shrimps; an encrusting coral (*Porites* in a new guise); many chitons, some new; and several octopi. The most obvious animals were the same marine pulmonates we had found at Angeles Bay, and these must have been strong and tough, for they were in the high rocks, fairly dry and exposed to the killing sun. The rocky ledge was covered with barnacles. The change in animal sizes on different levels was interesting. In the high-up pools there were small animals, mussels, snails, hermits, limpets, barnacles, sponges; while in the lower pools, the same species were larger. Among the small rocks and coarse gravel we found a great many stinging worms and a type of ophiuran new to us—actually it turned out to be the familiar *Ophionereis* in its juvenile stage. These high tide pools can be regarded as nurseries for more submerged zones. There were urchins, both club- and sharp-spined, and, in the sand, a few heart-urchins. The caverns under the rocks, exposed by the receding tide, were beautiful with many species of sponge, some pure white, some blue, and some purple, encrusting the rock surface. These under-rock caverns were as beautiful as those near Point Lobos in Central California. It was a long job to lay out and list the animals taken; meanwhile the crab-nets meant for the bottom were straining the current. In them we caught a number of very short fat stinging worms (*Chloeia viridis*), a species we had not seen before, probably a deep-water form torn loose by the strength of the current. With a hand-net we took a pelagic nudibranch, *Chioraera leonina,* found also in Puget Sound. The water swirled past the boat at about four miles an hour and we kept the dip-nets out until late at night. This was a strange collecting place. The water was quite cold, and many of the members of both the northern and the southern fauna occurred here. In this harbor there were conditions of stress, current, waves, and cold which seemed to encourage animal life. And it is reasonable that this should be so, for active, churning water means not only a strong oxygen content, but the constant movement of food. And in addition, the very difficulties involved in such a position—necessity for secure footing, crowding, and competition—seem to encourage a ferocity and a tenacity in the animals which go past survival and into successful reproduction. Where there is little danger, there seems to be little stimulation. Perhaps the pattern of struggle is so deeply imprinted in the genes of all life conceived in this benevolently hostile planet that the removal of obstacles automatically atrophies a survival drive. With warm water and abundant food, the animals may retire into a sterile sluggish happiness. This has certainly seemed true in man. Force and cleverness and versatility have surely been the children of obstacles. Tacitus, in the *Histories,* places as one of the tactical methods advanced to be used against the German armies their exposure to a warm climate and a soft rich food supply. These, he said, will ruin troops quicker than anything else. If these things are true in a biologic sense, what is to become of the fed, warm, protected citizenry of the ideal future state?

The classic example of the effect of such protection on troops is that they invariably lost discipline and wasted their energies in weak quarrelsomeness. They were never happy, never contented, but always ready to indulge in bitter and bloody personal quarrels. Perhaps this has no emphasis. So far there has been only one state that we know of which protected its people *without* keeping them constantly alert and organized against a real or imaginary outside enemy. This was the pre-Pizarro Inca state, whose people were so weakened that a little band of fierce, hard-bitten men was able to overcome the whole nation. And of them the converse is also true. When the food supply was wasted and destroyed by the Spaniards, when the fine economy which had distributed clothing and grain was overturned, only then, in their hunger and cold and misery, did the Peruvian people become a dangerous striking force. We have little doubt that a victorious collectivist state would collapse only a little less quickly than a defeated one. In fact, a bitter defeat would probably keep a fierce conquest-ideal alive much longer than a victory, for men can fight an enemy much more successfully than themselves.

Islands have always been fascinating places. The old story-tellers, wishing to recount a prodigy, almost invariably fixed the scene on an island—Faëry and Avalon, Atlantis and Cipango, all golden islands just over the horizon where anything at all might happen. And in old days at least it was rather difficult to check up on them. Perhaps this quality of potential prodigy still lives in our attitude toward islands. We want very much to go back to Guardian Angel with time and supplies. We wish to go over the burned hills and snake-ridden valleys, exposed to heat and insects, venom and thirst, and we are willing to believe almost anything we hear about it. We believe that great gold nuggets are found there, that unearthly animals make their homes there, that the mountain sheep, which is said never to drink water, abounds there. And if we were told of a race of troglodytes in possession, we should think twice before disbelieving. It is one of the golden islands which will one day be toppled by a mining company or a prison camp.

Thus far, there had been no illness on board the *Western Flyer*. Tiny drooped a little at Puerto Refugio and confessed that he didn't feel very well, and we held a consultation on him in the galley, explaining to him that consultation was more pleasant to us, as well as to him, than autopsy. After a great many questions, some of which might have been considered personal but which Tiny used as a vehicle for outrageous boasting, we concocted a remedy which might have cured almost anything—which was apparently what he had. Tiny emerged on deck some hours later, shaken but smiling.

He said that what he had been considering love had turned out to be simple flatulence. He said he wished all his romantic problems could be solved as easily.

It was now a long time since Sparky and Tiny had been able to carry out the good-will they felt toward Mexico, and they grew a little anxious about getting to Guaymas. There was no actual complaining, but they spoke tenderly of their intentions. Tex was inhibited in his good-will by his engagement to be married, which he wouldn't mention any more for fear we would diet him again. As for Tony, the master, he had no nerves, but the problems of finding new and unknown anchorages seemed to fascinate him. Tony would have made a great exploration captain. There would be few errors in judgment where he was concerned. The others of us were very busy all the time. We mention the health of the crew because we truly believe that the physical condition, and through it the mind, has reins on the actual collecting of animals. A man with a sore finger may not lift the rock under which an animal lives. We are likely to see more through our indigestion than through our eyes, and it seems to us that the ulcer-warped viewpoint is very often evident in animal descriptions. The man best fitted to observe animals, to understand them emotionally as well as intellectually, would be a hungry and libidinous man, for he and the animals would have the same preoccupations. Perhaps we fulfilled these requirements as well as most.

24

APRIL 3

We sailed around the northern tip of Guardian Angel and down its eastern coast. The water was clear and blue, and a large swell flowed past us. About noon we moved through a great group of Zeppelin-shaped jellyfish, ctenophores or possibly siphonophores. They were six to ten inches long, and the sea was littered with them. We slowed down and tried to scoop them up, but the tension of their bodies was not sufficient to hold them together out of water. They broke up and slithered in pieces through the dip-nets. Soon after, a school of whales went by, one of them so close that the spray from his blow-hole came over our deck. There is nothing so evil-smelling as a whale anyway, and a whale's breath is frightfully sickening. It smells of complete decay. Perhaps the droplets were left on the boat, for it seemed to us that we could smell him for a long time after he had gone by. The great schools of tuna, so evident in the Lower Gulf, were not seen here, but a few seals lazed through the water, and on one or two occasions we nearly ran over one asleep on the surface. We felt deeply the loneliness of this sea; no ships, no boats, no canoes, no little ranches on the shore nor villages. Now we would have welcomed a fishing Indian to come aboard and eat canned fruit salad, but this is a deserted sea.

The queer shoulder of Tiburón showed to the southeast of us, and we ran down on it with the wind behind us and probably the tidal current too, for we made great speed. We went down the western coast of Tiburón and watched its high cliffs through the glasses. The cliffs are fairly sheer, and the mountains are higher than those on Guardian Angel Island. This is the island where the Seri Indians come during parts of the year. It is said of them that they are or have been cannibals, a story which has been firmly denied again and again. It is certain that they have killed many strangers, but whether or not they have eaten them does not seem to be documented. Cannibalism is a fascinating subject to most people, and in some way a sin. Possibly the deep feeling is that if people learn to eat one another the food supply would be so generous and so available that no one would be either safe or hungry. It is very curious the amount of hatred and fear that cannibalism inspires. These poor Seri Indians would not be so much feared for their murdering habits, but if in their hunger they should cut a steak from an American citizen a panic arises. Swift's quite reasonable suggestion concerning a possible use for Irish babies aroused a storm of emotionalism out of all proportion to its feasibility. There were not, it is thought, enough well-conditioned babies at that time to have provided anything like an adequate food supply. Swift without a doubt meant it only as an experiment. If it had been successful, there would have been time enough then to think of raising more babies. It has generally been found that starvation is the greatest single cause for cannibalism. In other words, people will not eat each other if they can get anything else. To some extent this reluctance must be caused by an unpleasant taste in human flesh, the result no doubt of our rather filthy eating habits. This need not be a future deterrent, for, if other barriers are removed, such as a natural distaste for eating relatives, or a man's gallant dislike for eating women, who in turn are inhibited by a romantic tendency—if all these difficulties should be solved it would be easy enough to improve the flavor of human flesh by special diets before slaughter and carefully prepared sauces and condiments afterwards. If this should occur, the Seri Indians, if indeed they do be cannibals, far from being loaded with our hatred, must be considered pioneers in a new field and honored as such.

Clavigero, in his *History of* [Lower] *California*,[1] has an account of these Seris.

> The vessel, *San Javier* [he says], which had left Loreto in September 1709 with three thousand *scudi* to buy provisions in Yaqui, was carried 180 miles above the port of its destination by a furious storm and grounded on the sand. Some of the people were drowned; the rest saved themselves in the small boat; but after landing they were exposed to another not less serious danger because that coast was inhabited by the Series who were warlike gentiles and implacable enemies of the Spaniards. For this reason they hastened to bury the

[1] Lake and Gray translation, 1937, pp. 217–18.

money and all the possessions which were on the boat; and after embarking again in the small boat they continued with a thousand dangers and hardships to Yaqui, from where they sent the news to Loreto. In a little while the Serìes came to the place where the Spaniards had buried those possessions, and they dug them up and carried them away. They even removed the rudder from the vessel and they destroyed it in order to get the nails.

As soon as Father Salvatierra learned of that misfortune, he left in the unseaworthy vessel, the *Rosario,* and went to the port of Guaymas. From there he sent this vessel to the place where the *San Javier* was grounded, and he himself went with fourteen Yaqui Indians in that direction over a very bad road which absolutely lacked potable water, and for this reason they suffered great thirst for two days. During the two months which he lived there, exposed to hunger and hardships and to the great danger of all their lives (while the vessel was being repaired), he won the good-will of the Serìes in such manner that he not only recovered all the cargo of the boat which they had stolen but induced them also to make peace with the Pimas, who were Christian neighbors of theirs and enemies whom they most hated. He baptized many of their children, he catechized the adults and inspired so much affection in them for Christianity that they immediately wanted a missionary to instruct them regularly and to baptize them and govern them in all respects.

So the dominating sweetness of the character of Father Salvatierra, aided by the grace of the Master, triumphed over the ferocity of those barbarians who were so feared, not only by the other Indians, but also by the Spaniards. He wept tenderly on seeing their unexpected docility and their good inclinations, thanking God for having had that much good come from the misfortune of the vessel.

The "dominating sweetness" of the character of Father Salvatierra did not, however, change them completely, for they have gone right on killing people until recently. In this account it is also interesting to notice Clavigero's statement that Father Salvatierra took the "unseaworthy" *Rosario.* In the long record of wrecks blowing off course, of marine disaster of every kind, it was wonderful how they were able to judge whether or not a ship was unseaworthy. A little reading of contemporary records of voyaging by these priestly and soldierly navigators indicates that they put more faith in prayer than in the compass. We think the present-day navigators of the Gulf have learned their seamanship in the same school. Some of the ships we saw at Guaymas and La Paz floated in violation of every law of physics. There must be in Heaven a small pilot-house where a worried and distraught St. Christopher spends a good deal of his time looking after the shipping of the Gulf of California with a handful of miracles.

Tiburón looked red to us, and the brush seemed stronger and greener than any we had seen in a long time. In some of the creases between the hills there were growths of small ground-hugging trees like our scrub-oaks. What they were, of course, we do not know. About five-thirty in the afternoon we rounded Red Bluff Point on the southwesterly corner of Tiburón and came to anchor in the lee of the long point, protected from northerly winds. The "corner" of the island is a chosen word, for Tiburón is a rough square lying plumb with the points of the compass. We searched the shore for Seris and saw none. In our usual condition of hunger, it would have been a toss-up whether Seris ate us or we ate Seris. The one who got in the first bite would have had the dinner, but we never did see a Seri.

The coast at this station was interesting; off Red Bluff Point low flat rocks shelved gradually seaward—fine collecting rocks, with many of the pot-holes which make such beautiful natural aquaria at low tide. Southward of this were long reef-like stone fingers extending outward, with shallow sand-bottom baylets between them, almost like boat slips. Next to this was a bouldery beach with stones imbedded in sand; and finally a coarse sand beach. Here again was nearly every kind of environment except mud-flat and lagoon. We began our collecting on the reef, and found the little pot-holes lovely with hydroids and coral and colored sponge and little bright algae. There were many broken-back shrimps in these pools, difficult little fellows to catch, for they are so transparent as to be almost invisible and they move with great speed by flipping their tails like lobsters. Only their stomachs and flickering gills are visible, and one can watch their insides work as though they were little glass models. We caught many of them by working our hands very slowly under them and raising them gradually to the surface.

On the reef, there were the usual *Heliasters,* anemones, and cucumbers, urchins, and a great number of giant snails,[2] of which we

[2] Callopoma fluctuosum.

collected many hundreds. High up in the intertidal were many *Tegula*-like snails of the kind we found at Cape San Lucas, although here the water was clear and very cold whereas at Cape San Lucas it had been warm. There were very few Sally Lightfoots here; *Pachygrapsus,* the northern crabs, had taken their place. We took abundant solitary corals and laid in a large supply of plumularian hydroids, gathered carefully and preserved so that they might not be crushed or broken. These animals, in appearance at least, are so like plants that they indicate to the imagination a bridge between flora and fauna, just as some plants, like the tropical sensitive plants and the insect-eating plants, indicate by their apparent nervous and muscular versatility an approach from the other side.

On the reef, we took a number of barnacles, many *Phataria* and *Linckia,* sponges, and tunicates. Moving from the reef to the stone fingers, we saw and captured a most attenuated spider crab,[3] all legs and little body. On the sand bottom between the fingers were many sting-rays lying quietly, and near the edge of one little harbor there were two in copulation, male (or female) lying on its back with its mate on top of it and the heads together. We wanted these two, and so after a moment in which we toughened the fibers of our romantic feelings, we put a light harpoon through both of them at once and brought them up, angry and disillusioned. We had hoped that they might remain fastened so that they might be preserved in coition, but their softer feelings were offended and they disengaged.

Meanwhile Tiny, moving in the little slip-like bays with the skiff, harpooned several more sting-rays. On the beach we took several sand-living cucumbers, and in the bottom of a mud pool searched long and unsuccessfully for a furry crab which had been seen scuttling into its burrow. This was a rich field for collecting, but the horizon markers were true to their position in the Gulf, and except for the profusion on Red Bluff Point reef, where the footing was excellent, there was nothing novel.

When it grew dark, we turned on the deck lights and saw numbers of a barracuda-like fish coming to eat the small fishes that gathered to the light. We put a fish-line on a small trident spear and began throwing it at them. About every tenth cast we struck one and brought him to the deck. And now a curious thing hap-

[3] Stenorhynchus debilis.

pened. From the shore came a swarm of very large bats. Their bodies were small but they had a twelve- to fifteen-inch wingspread. They circled restlessly around the boat, although there were no insects about. Sparky was on the rail, spearing barracuda, and he is very much afraid of bats. Suddenly one swooped near him, and he struck at it with the harpoon. By one of those strange accidents, the barbs went into the bat and captured it, and now four or five more dived straight at Sparky's head and he dropped the harpoon and ran for the galley. The dead bat fell over the side into the water, where we later picked it up.

Then an even stranger thing happened. As though at a signal, every bat of the hundreds suddenly turned and flew away to shore and not another one was seen. We have not yet a report on the one taken, so we do not know what kind of bats they were. There are reports of fish-eating bats, and these may have been that kind. We warned Sparky seriously to keep very quiet about the incident. "Sparky," we said, "we know that your reputation for truthfulness in Monterey is as good as most. In other words, it is not above reproach. If we were you, when you get back to Monterey, we would never mention to anyone that we had harpooned a bat. We would make up stories and adventures, but there is no reason for straining an already shaky reputation." Sparky promised he would never tell, but back in Monterey he couldn't resist and, just as we supposed, a roar of laughter went up. In Monterey they said, "You know what that Sparky said? He swears he harpooned a bat."

And as punishment to Sparky, when we were questioned we said, "Bat? What bat?" Sparky is a little touchy about the whole subject, and he dislikes bats very intensely now.

Meanwhile we had twelve of the barracuda-like fish. We preserved some of them but did not try to eat any. The sierras and tuna were too delicious to justify making experiments with strong fish.

The mountains of Tiburón were very black against the stars and the sea was calm. On the deck, Tiny made a little noise washing a shirt, for we were not far from Guaymas and Tiny was growing anxious. We discussed bats, and the horror they create in people and the myths about them—in his *Caribbean Treasure*, page 56, Ivan Sanderson makes some very interesting remarks about vampire bats as carriers of rabies, and their whole tie-in with the vampire tradition, so intimately related to werewolfism in the popular mind. A man with rabies, one might infer, could well be the werewolf which occurs all over the world, and vampire and werewolf very often go together. It is a fascinating speculation, and surely the unreasoning and almost instinctive fear of bats might indicate another of those memory-like patterns, some horrible recollection of the evil bats can do.

We find after reading many scientific and semi-scientific accounts of exploration that we have two strong prejudices: the first of these arises where there is a woman aboard—the wife of one of the members of the party. She is never called by her name or referred to as an equal. In the account she emerges as "the shipmate," the "skipper," the "pal." She is nearly always a stringy blonde with leathery skin who is included in all photographs to give them "interest." Our second prejudice concerns a hysteria of love which manifests itself in an outcry against parting and is usually written in Spanish. This outburst comes at the end of the book. It goes, "And so——." Always, "and so," for some reason. "And so we said good-by to Tiburón, vowing to come back again. *Adiós, Tiburón, amigo,* friend." For some reason this stringy shipmate and this rush of emotion are slightly obscene to us. And so we said good-by to Tiburón and trucked on down toward Guaymas.

25

APRIL 22

The trolling jigs picked up two fine sierras on the way. Our squid jigs had gone to pieces from much use, and had to be repaired with white chicken feathers. We were under way all day, and toward evening began to see the sport-fishing boats of Guaymas with their cargoes of sportsmen outfitted with equipment to startle the fish into submission. And the sportsmen were mentally on tiptoe to out-think the fish—which they sometimes do. We thought it might be fun some time to engage in this intellectual approach toward fishing, instead of our barbarous method of throwing a line with a chicken-feather jig overboard. These fishermen in their swivel fishing chairs looked comfortable and clean and pink. We had been washing our clothes in salt water, and we felt sticky and salt-crusted; and, being less comfortable and clean than the sportsmen, we built a whole defense of contempt. With no effort at all on their part we had a good deal of dislike for them. It is probable that Sparky and Tiny had a true contempt, uncolored by envy, for they are descended from many generations of fishermen who went out for fish, not splendor. But even they might have liked sitting in a swivel chair holding a rod in one hand and a frosty glass in the other, blaming a poor day on the Democrats, and offering up prayers for good fishing to Calvin Coolidge.

We could not run for Guaymas that night, for the pilot fees rise after hours and we were getting a little low on money. Instead, about six P.M. we rounded Punta Doble and put into Puerto San Carlos. This is another of those perfect little harbors with narrow rocky entrances. The entrance is less than eight hundred yards wide, and steep rocks guard it. Once inside, there is anchorage from five to seven fathoms. The head of the bay is bordered by a sand beach, changing to boulders near the entrance. There was still time for collecting.

We went to the bouldery beach and took some snails new to us and two echiuroids. But nothing on or under the rocks was different from the Tiburón animals. The water was warm here and it was soupy with shrimps, of which we took a number in a dip-net. We made a quick survey of the area, for darkness was coming. As soon as it was dark we began to hear strange sounds in the water around the *Western Flyer*—a periodic hissing and many loud splashes. We went to the deckhouse and turned on the searchlight. The bay was swarming with small fish, apparently come to eat the shrimps. Now and then a school of six- to ten-inch fish would drive at the little fish with such speed and in such numbers that they made the sharp hissing we had heard, while farther off some kind of great fish leaped and splashed heavily. Without a word, Sparky and Tiny got out a long net, climbed into the skiff, and tried to draw their net around a school of fish. We shouted at them, asking what they would do with the fish if they caught them, but they were deaf to us. The numbers of fish had set off a passion in them—they were fishermen and the sons of fishermen—let businessmen dispose of the fish; their job was to catch them. They worked frantically, but they could not encircle a school, and soon came back exhausted.

Meanwhile, the water seemed almost solid with tiny fish, one and one-half to two inches long. Sparky went to the galley and put the biggest frying pan on the fire and poured olive oil into it. When the pan was very hot he began catching the tiny fish with the dip-nets, a hundred or so in each net. We passed the nets through the galley window and Sparky dumped them into the frying pan. In a short time these tiny fish were crisp and brown. We drained, salted, and ate them without any cleaning at all and they were delicious. Probably no fresher fish were ever eaten, except perhaps by the Japanese, who are said to eat them alive, and by college boys, who are photographed doing it. Each fish was a curled, brown, crisp little bite, delicate and good. We ate hundreds of them. Afterwards we went back to the usual night practice of netting the pelagic animals which came to the light. We took shrimps and larval shrimps, numbers of small swimming crabs, and more of the transparent fish. All night the hissing rush and splash of hunters and hunted went on. We had never been in water so heavily populated. The light, piercing the surface, showed the water almost solid with fish—swarming, hungry, frantic fish, incredible in their voraciousness. The schools swam, marshaled and patrolled. They turned as a unit and dived as a unit. In their millions they followed a pattern minute as to direction and depth and speed. There must be some fallacy in our thinking of these fish as individuals. Their functions in the school are in some as yet unknown way as controlled as though the school were one unit. We cannot conceive of this intricacy until we are able to think of the school as an animal itself, reacting with all its cells to stimuli which perhaps might not influence one fish at all. And this larger animal, the school, seems to have a nature and drive and ends of its own. It is more than and different from the sum of its units. If we can think in this way, it will not seem so unbelievable that every fish heads in the same direction, that the water interval between fish and fish is identical with all the units, and that it seems to be directed by a school intelligence. If it is a unit animal itself, why should it not so react? Perhaps this is the wildest of speculations, but we suspect that when the school is studied as an animal rather than as a sum of unit fish, it will be found that certain units are assigned special functions to perform; that weaker or slower units may even take their places as placating food for the predators for the sake of the security of the school as an animal. In the little Bay of San Carlos, where there were many schools of a number of species, there was even a feeling (and "feeling" is used advisedly) of a larger unit which was the inter-relation of species with their interdependence for food, even though that food be each other. A smoothly working larger animal surviving within itself—larval shrimp to little fish to larger fish to giant fish—one operating mechanism. And perhaps *this* unit of survival may key into the larger animal which is the life of all the sea, and this into the larger of the world. There would seem to be only one commandment for living things: Survive! And the forms and species and units and groups are armed for survival, fanged for survival, timid for it, fierce for it, clever for it, poisonous for it, intelligent for it. This commandment decrees the death and destruction of myriads of individuals for the survival of the whole. Life has one final end, to be alive; and all the tricks and mechanisms, all the successes and all the failures, are aimed at that end.

26

APRIL 5

We sailed in the morning on the short trip to Guaymas. It was the first stop in a town that had anything like communication since we had left San Diego. The world and the war had become remote to us; all the immediacies of our usual lives had slowed up. Far from welcoming a return, we rather resented going back to newspapers and telegrams and business. We had been drifting in some kind of dual world—a parallel realistic world; and the preoccupations of the world we came from, which are considered realistic, were to us filled with mental mirage. Modern economies; war drives; party affiliations and lines; hatreds, political, and social and racial, cannot survive in dignity the perspective of distance. We could understand, because we could feel, how the Indians of the Gulf, hearing about the great ant-doings of the north, might shake their heads sadly and say, "But it is crazy. It would be nice to have new Ford cars and running water, but not at the cost of insanity." And in us the factor of time had changed: the low tides were our clock and the throbbing engine our second hand.

Now, approaching Guaymas, we were approaching an end. We planned only two or three collecting stations beyond, and then the time of charter-end would be crowding us, and we would have to run for it to be back when the paper said we would. The charter at least fixed our place in time. And already our crew was trying to think of ways to come back to the Gulf. This trip had been like a dreaming sleep, a rest from immediacies. And in our contacts with Mexican people we had been faced with a change in expediencies. Perhaps—even surely—these people are expedient, but on some other plane than our ordinary one. What they did for us was without hope or plan of profit. We suppose there must have been some kind of profit involved, but not the kind we are used to, not of material things changing hands. And yet some trade took place at every contact—something was exchanged, some unnamable of great value. Perhaps these people are expedient in the unnamables. Maybe they bargain in feelings, in pleasures, even in simple contacts. When the Indians came to the *Western Flyer* and sat timelessly on the rail, perhaps they were taking something. We gave them presents, but it is sure they had not come for presents. When they helped us, it was with no idea of material payment. There were material prices for material things, but one couldn't buy kindness with money, as one can in our country. It was so in every contact, and they were so used to the spiritual transaction that they had difficulty translating material things into money. If we wanted to buy a harpoon, there was difficulty immediately. What was the price? An Indian had paid three pesos for the harpoon several years ago. Obviously, since that had been paid, that was the price. But he had not yet learned to give time a money value. If he had to go three days in a canoe to get another harpoon, he could not add his time to the price, because he had never thought of time as a medium of exchange. At first we tried to explain the feeling we all had that time is a salable article, but we had to give it up. Time, these Indians said, went on. If one could stop time, or take it away, or hoard it, then one might sell it. One might as well sell air or heat or cold or health or beauty. And we thought of the great businesses in our country—the sale of clean air, of heat and cold, the scrabbling bargains in health offered over the radio, the boxed and bottled beauty, all for a price. This was not bad or good, it was only different. Time and beauty, they thought, could not be captured and sold, and we knew they not only *could* be, but that time could be warped and beauty made ugly. And again it was not good or bad. Our people would pay more for pills in a yellow box than in a white box—even the refraction of light had its price. They would buy books because they should rather than because they wanted to. They bought immunity from fear in salves to go under their arms. They bought romantic adventure in bars of tomato-colored soaps. They bought education by the foot and hefted the volumes to see that they were not short-weighted. They purchased pain, and then analgesics to put down the pain. They bought courage and rest and had neither. And they are vastly amused at the Indian who, with his silver, bought Heaven and ransomed his father from Hell. These Indians were far too ignorant to understand the absurdities merchandising can really achieve when it has an enlightened people to work on.

One can go from race to race. It is coming back that has its violation. As we feel greatness, we feel that these people are very great. It seems to us that the repose of an Indian woman sitting in the gutter is beyond our achievement. But even these people wish for our involvement in temporal and material things. Once we thought that the bridge between cultures might be through education, public health, good housing, and through political vehicles —democracy, Nazism, communism—but now it seems much simpler than that. The invasion comes with good roads and high-tension wires. Where those two go, the change takes place very quickly. Any of the political forms can come in once the radio is hooked up, once the concrete highway irons out the mountains and destroys the "localness" of a community. Once the Gulf people are available to contact, they too will come to consider clean feet more important than clean minds. These are the factors of civilization and their paths, good roads, high-voltage wires, and possibly canned foods. A local 110-volt power unit and a winding dirt road may leave a people for a long time untouched, but high-voltage operating day and night, the network of wires, will draw the people into the civilizing web, whether it be in Asiatic Russia, in rural England, or in Mexico. That *Zeitgeist* operates everywhere, and there is no escape from it.

Again, this is not to be considered good or bad. To us, a little weary of the complication and senselessness of a familiar picture, the Indian seems a rested, simple man. If we should permit ourselves to remain in ignorance of his complications, then we might long for his condition, thinking it superior to ours. The Indian on the other hand, subject to constant hunger and cold, mourning a grandfather and set of uncles in Purgatory, pained by the aching teeth and sore eyes of malnutrition, may well envy us our luxury. It is easy to remember how, when we were in the terrible complication of childhood, we longed for easy and uncomplicated adulthood. Then we would have only to reach into our pockets for money, then all problems would be ironed out. The ranch-owner had said, "There is no poverty in your country and no misery. Everyone has a Ford."

We arrived early at Guaymas, passed the usual tests of customs, got our mail at the consulate, and then did the various things of the port. Some of those things are amusing, but they are out of drawing for this account. Guaymas was already in the pathway of the good highway; it was no longer "local." At La Paz and Loreto the Gulf and the town were one, inextricably bound together, but here at Guaymas the railroad and the hotel had broken open that relationship. There were gimcracks for tourists everywhere. This is no criticism of the change, but Guaymas seems to us to be outside the boundaries of the Gulf. We had good treatment there, met charming people, did good and bad things, and left with reluctance.

27

APRIL 8

We sailed out on Monday, a little tattered and a little tired. Captain Corona, pilot and shrimp-boat owner, who had been kind and hospitable to us, piloted us out and stopped one of his incoming boats for us to inspect. It was a poor small boat, and had not much of a catch of shrimps. Everyone in this neighborhood had complained of the Japanese shrimpers who were destroying the shrimp fisheries. We determined to pay them a visit on the next day. The moment we dropped the pilot, just outside of Guaymas, the Gulf was local again and part of the design it had put in our heads. The mirage was over the land and the sea was very blue. We sailed only a short distance and dropped our anchor in a little cove opposite the Pajaro Island light. That night we caught a number of fish that looked and felt like catfish. Tex skinned them and prepared them, and we did not eat them. A little gloom hung over all of us; Sparky and Tiny had fallen in love with Guaymas and planned to go back there and live forever. But Tex and Tony were gloomy and a little homesick.

We were awakened well before daylight by the voices of men paddling out for the day's fishing, and it was with some relief that we pulled up our anchor and started out to continue the work we had come to do. The day was thick with haze, the sun came through it hot and unpleasant, and the water was oily and at the same time choppy. The sticky humidity was on us.

In about an hour we came to the Japanese fishing fleet. There were six ships doing the actual dredging while a large mother ship of at least 10,000 tons stood farther offshore at anchor. The dredge boats themselves were large, 150 to 175 feet, probably about 600 tons. There were twelve boats in the combined fleet including the mother ship, and they were doing a very systematic job, not only of taking every shrimp from the bottom, but every other living thing as well. They cruised slowly along in echelon with overlapping dredges, literally scraping the bottom clean. Any animal which escaped must have been very fast indeed, for not even the sharks got away. Why the Mexican government should have permitted the complete destruction of a valuable food supply is one of those mysteries which have their ramifications possibly back in pockets it is not well to look into.

We wished to go aboard one of the dredge boats. Tony put the *Western Flyer* ahead of one of them, and we dropped the skiff over the side and got into it. It was not a friendly crew that looked at us over the side of the iron dredge boat. We clung to the side, almost swamping the skiff, and passed our letter from the Ministry of Marine aboard. Then we hung on and waited. We could see the Mexican official on the bridge reading our letter. And then suddenly the atmosphere changed to one of extreme friendliness. We were helped aboard and our skiff was tied alongside.

The cutting deck was forward, and the great dredge loads were dumped on this deck. Along one side there was a long cutting table where the shrimps were beheaded and dropped into a chute, whether to be immediately iced or canned, we do not know. But probably they were canned on the mother ship. The dredge was out when we came aboard, but soon the cable drums began to turn, bringing in the heavy purse-dredge. The big scraper closed like a sack as it came up, and finally it deposited many tons of animals on the deck—tons of shrimps, but also tons of fish of many varieties: sierras; pompano of several species; of the sharks, smooth-hounds and hammer-heads; eagle rays and butterfly rays; small tuna; catfish; *puerco*—tons of them. And there were bottom-samples with anemones and grass-like gorgonians. The sea bottom must have been scraped completely clean. The moment the net dropped open and spilled this mass of living things on the deck, the crew of Japanese went to work. Fish were thrown overboard immediately, and only the shrimps kept. The sea was littered with dead fish, and the gulls swarmed about eating them. Nearly all the fish were in a dying condition, and only a few recovered. The waste of this good food supply was appalling, and it was strange that the Japanese, who are usually so saving, should have done it. The shrimps were shoveled into baskets and delivered to the cutting table. Meanwhile the dredge had gone back to work.

With the captain's permission, we picked out several representatives of every fish and animal we saw. A stay of several days on the boat would have been the basis of a great and complete collection of every animal living at this depth. Even going over two dredgeloads gave us many species. The crew, part Mexican and part Japanese, felt so much better about us by now that they brought out their treasures and gave them to us: bright-red sea-horses and brilliant sea-fans and giant shrimps. They presented them to us, the rarities, the curios which had caught their attention.

At intervals a high, chanting cry arose from the side of the ship and was taken up and chanted back from the bridge. From the upper deck a slung cat-walk extended, and on it the leadsman stood, swinging his leadline, bringing it up and swinging it out again. And every time he read the markers he chanted the depth in Japanese in a high falsetto, and his cry was repeated by the helmsman.

We went up on the bridge, and as we passed this leadsman he said, "Hello." We stopped and talked to him a few moments before we realized that that was the only English word he knew. The Japanese captain was formal, but very courteous. He spoke neither Spanish nor English; his business must all have been done through an interpreter. The Mexican fish and game official stationed aboard was a pleasant man, but he said that he had no great information about the animals he was overseeing. The large shrimps were *Penaeus stylirostris,* and one small specimen was *P. californiensis.*

The shrimps inspected all had the ovaries distended and apparently, as with the Canadian *Pandalus,* this shrimp had the male-female succession. That is, all the animals are born male, but all become females on passing a certain age. The fish and game man seemed very eager to know more about his field, and we promised to send him Schmitt's fine volume on *Marine Decapod Crustacea of California* and whatever other publications on shrimps we could find.

We liked the people on this boat very much. They were good men, but they were caught in a large destructive machine, good men doing a bad thing. With their many and large boats, with their industry and efficiency, but most of all with their intense energy, these Japanese will obviously soon clean out the shrimps of the region. And it is not true that a species thus attacked comes back. The disturbed balance often gives a new species ascendancy and destroys forever the old relationship.

In addition to the shrimps, these boats kill and waste many hundred of tons of fish every day, a great deal of which is sorely needed for food. Perhaps the Ministry of Marine had not realized at that time that one of the good and strong food resources of Mexico was being depleted. If it has not already been done, catch limits should be imposed, and it should not be permitted that the region be so intensely combed. Among other things, the careful study of this area should be undertaken so that its potential could be understood and the catch maintained in balance with the supply. Then there might be shrimps available indefinitely. If this is not done, a very short time will see the end of the shrimp industry in Mexico.

We in the United States have done so much to destroy our own resources, our timber, our land, our fishes, that we should be taken as a horrible example and our methods avoided by any government and people enlightened enough to envision a continuing economy. With our own resources we have been prodigal, and our country will not soon lose the scars of our grasping stupidity. But here, with the shrimp industry, we see a conflict of nations, of ideologies, and of organisms. The units of the organisms are good people. Perhaps we might find a parallel in a moving-picture company such as Metro-Goldwyn-Mayer. The units are superb—great craftsmen, fine directors, the best actors in the profession—and yet due to some overlying expediency, some impure or decaying quality, the product of these good units is sometimes vicious, sometimes stupid, sometimes inept, and never as good as the men who make it. The Mexican official and the Japanese captain were both good men, but by their association in a project directed honestly or dishonestly by forces behind and above them, they were committing a true crime against nature and against the immediate welfare of Mexico and the eventual welfare of the whole human species.

The crew helped us back into our skiff, handed our buckets of specimens down to us, and cast us off. And Tony, who had been cruising slowly about, picked us up in the *Western Flyer*. We had taken perhaps a dozen pompano as specimens when hundreds were available. Sparky was speechless with rage that we had brought none back to eat, but we had forgotten that. We set our course southward toward Estero de la Luna—a great inland sea, the borders of which were dotted lines on our maps. Here we expected to find a rich estuary fauna. In the scoop between Cape Arco and Point Lobos there is a fairly shallow sea which makes a deep ground-swell. It was Tiny up forward who noticed the great numbers of manta rays and suggested that we hunt them. They were monsters, sometimes twelve feet between the "wing" tips. We had no proper equipment, but finally we rigged one of the arrow-tipped harpoons on a light line. This harpoon was a five-inch bronze arrow slotted on an iron shaft. After the stroke, the arrow turns sideways in the flesh and the shaft comes out and floats on its wooden handle. The line is fastened to the arrow itself.

The huge rays cruised slowly about, the upturned tips of the wings out of water. Sparky went to the crow's-nest, where he could look down into the water and direct the steersman. A hundred feet from one of the great fish we cut the engine and coasted down on it. It lay still on the water. Tiny poised prettily on the bow. When we were right on it, he drove the harpoon into it. The monster did not flurry, it simply faded for the bottom. The line whistled out to its limit, twanged like a violin string, and parted. A curious excitement ran through the boat. Tex came down and brought out a one-and-one-half-inch hemp line. He ringed this into a new harpoon-head, and again Tiny took up his position, so excited that he had his foot in a bight of the line. Luckily, we noticed this and warned him. We coasted up to another ray—Tiny missed the stroke. Another, and he missed again. The third time, his arrow drove home, the line sang out again, two hundred feet of it. Then it came to the end where it was looped over a bitt, vibrated for a moment, and parted. The breaking strain of this rope was enormous. But we were doing it all wrong and we knew it. A ton and a half of speeding fish is not to be brought up short. We should have thrown a keg overboard with the line and let the fish fight the keg's buoyancy until exhausted. But we were not equipped. Tiny was almost hysterical by this time. Tex brought a three-inch line with an extremely high breaking strain. We had no more arrow harpoons. Tex made his hawser fast to a huge trident spear. When he finished, the assembly was so heavy that one man could hardly lift it. This time, Tex took the harpoon. He did not waste his time with careless strokes; he waited until the bow was right over one of the largest rays, then drove his spear down with all his strength. The heavy hawser ran almost smoking over the rail. Then it came to the bitt and struck with a kind of groaning cry, quivered, and went limp. When we pulled the big harpoon aboard, there was a chunk of flesh on it. Tiny was heart-broken. The wind came up now and so ruffled the water that we could not see the coasting monsters any more. We tried to soothe Tiny.

"What could we do with one if we caught it?" we asked.

Tiny said, "I'd like to pull it up with the boom and hang it right over the hatch."

"But what could you do with it?"

"I'd hang it there," he said, "and I'd have my picture taken with it. They won't believe in Monterey I speared one unless I can prove it."

He mourned for a long time our lack of foresight in failing to bring manta ray equipment. Late into the night Tex worked, making with file and emery stone a new arrow harpoon, but one of great size. This he planned to use with his three-inch hawser. He said the rope would hold fifteen to twenty tons, and this arrow would not pull out. But he never had a chance to prove it; we did not see the rays in numbers any more.

We came to anchorage that night south of Lobos light and about five miles from the entrance to Estero de la Luna. In this shallow water Tony did not like to go closer for fear of stranding. It was a strange and frightening night, and no one knew why. The water was glassy again and the deck soaked with humidity. We had a curious feeling that a stranger was aboard, some presence not seen but felt, a dark-cloaked person who was with us. We were all nervous and irritable and frightened, but we could not find what frightened us. Tex worked on the Sea-Cow and got it to running perfectly, for we wanted to use it on the long run ashore in the morning. We had checked the tides in Guaymas; it was necessary to leave before daylight to get into the estuary for the low tide. In the night, one of us had a nightmare and shouted for help and the rest of us were sleepless. In the darkness of the early morning, only two of us got up. We dressed quietly and got our breakfast. The light on Lobos Point was flashing to the north of us. The decks were soaked with dew. Climbing down to the skiff, one of us fell and wrenched his leg. True to form, the Sea-Cow would not start. We set off rowing toward the barely visible shore, fixing our course by Lobos Light. A little feathery white shape drifted over the water and it was joined by another and another, and very soon a dense white fog covered us. The *Western Flyer* was lost and the shore blotted out. With the last flashing of the Lobos light we tried to judge the direction of the swell to steer by, and then the light was gone and we were cut off in this ominous glassy water. The air turned steel-gray with the dawn, and the fog was so thick that we could not see fifteen feet from the boat. We rowed on, remembering to quarter on the direction of the swells. And then we heard a little vicious hissing as of millions of snakes, and we both said, "It's the *cordonazo*." This is a quick fierce storm which has destroyed more ships than any other. The wind blows so that it clips the water. We were afraid for a moment, very much afraid, for in the fog, the *cordonazo* would drive our little skiff out into the Gulf and swamp it. We could see nothing and the hissing grew louder and had almost reached us.

It seems to be this way in a time of danger. A little chill of terror runs up the spine and a kind of nausea comes into the throat. And then that disappears into a kind of dull "what the hell" feeling. Perhaps this is the working of some mind-to-gland-to-body process. Perhaps some shock therapy takes control. But our fear was past now, and we braced ourselves to steady the boat against the impact of the expected wind. And at that moment the bow of the skiff grounded gently, for it was not wind at all that hissed, but little waves washing strongly over an exposed sand-bar. We climbed out, hauled the boat up, and sat for a moment on the beach. We had been badly frightened, there is no doubt of it. Even the sleepy dullness which follows the adrenal drunkenness was there. And while we sat there, the fog lifted, and in the morning light we could see the *Western Flyer* at anchor offshore, and we had landed only about a quarter of a mile from where we had intended. The sun broke clear now, and true to form when there was neither danger nor much work the Sea-Cow started easily and we rounded the sand-hill entrance of the big estuary. Now that the sun was up, we could see why there were dotted lines on the maps to indicate the borders of the *estero*. It was endless—there were no borders. The mirage shook the horizon and draped it with haze, distorted shapes, twisted mountains, and made even the bushes seem to hang in the air. Until every foot of such a shore is covered and measured, the shape and extent of these estuaries will not be known.

Inside the entrance of the estuary a big canoe was anchored and four Indians were coming ashore from the night's fishing. They were sullen and unsmiling, and they grunted when we spoke to them. In their boat they had great thick mullet-like fish, so large that it took two men to carry each fish. They must have weighed sixty to one hundred pounds each. These Indians carried the fish through an opening in the brush to a camp of which we could see only the smoke rising, and they were definitely unfriendly. It was the first experience of this kind we had had in the Gulf. It wasn't that they didn't like *us*—they didn't seem to like each other.

The tide was going down in the estuary, making a boiling current in the entrance. Biologically, the area seemed fairly sterile. There were numbers of small animals, several species of large snails and a number of small ones. There were burrowing anemones[1] with transparent, almost colorless tentacles spread out on the sand bottom. And there were the flower-like *Cerianthus* in sand-tubes everywhere. On the bottom were millions of minute sand dollars of a new type, brilliant light green and having holes and fairly elongate spines. Farther inside the estuary we took a number of small heart-urchins and a very few larger ones. On the sand bottoms there were large burrows, but dig as we would, we could never find the owners. They were either very quick or very deep, but even under water their burrows were always open and piles of debris lay about the entrances. Some large crustacean, we thought, possibly of the fiddler crab clan.

The commonest animal of all was the enteropneust, an "acorn-tongued" worm presumably about three feet long that we had found at San Lucas Cove and at Angeles Bay. There were hundreds of their sand-castings lying about. We were not convinced that with all our digging we had got the whole animal even once (and the specialist subsequently confirmed our opinion with regret!).

Deep in the estuary we took several large beautifully striped *Tivela*-like clams and a great number of flat pearly clams. There were hundreds of large hermit crabs in various large gastropod shells. We found a single long-armed sand-burrowing brittle-star which turned out to be *Ophiophragmus marginatus*, and our listing of it is the only report on record since Lütken, in Denmark, erected the species from Nicaraguan material nearly a hundred years ago. In the uneasy footing of the sand, every stick and large shell and rock was encrusted with barnacles; even one giant swimming crab carried a load of barnacles on his back.

The wind had been rising a little as we collected, rippling the water. We cruised about in the shallows trying to see the bottom. There were great numbers of sand sharks darting about, but the

[1] Harenactis.

bottom was clean and sterile and not at all as well populated as we would have supposed. The mirage grew more and more crazy. Perhaps these sullen Indians were bewildered in such an uncertain world where nothing half a mile away could maintain its shape or size, where the world floated and trembled and flowed in dream forms. And perhaps the reverse is true. Maybe these Indians dream of a hard sharp dependable world as an opposite of their daily vision.

We had not taken riches in this place. When the tide turned we started back for the *Western Flyer*. Perhaps the Sea-Cow too had been frightened that morning, for it ran steadily. But the tide was so strong that we had to help it with the oars or it would not have been able to hold its own against the current. It took two hours of oars and motor to get back to the *Western Flyer*.

We felt that this had not been a good nor a friendly place. Some quality of evil hung over it and infected us. We were not at all sorry to leave it. Everyone on board was quiet and uneasy until we pulled up the anchor and started south for Agiabampo, which was, we thought, to be our last collecting station.

It was curious about this Estero de la Luna. It had been a bad place—bad feelings, bad dreams, and little accidents. The look and feel of it were bad. It would be interesting to know whether others have found it so. We have thought how places are able to evoke moods, how color and line in a picture may capture and warp us to a pattern the painter intended. If to color and line in accidental juxtaposition there should be added odor and temperature and all these in some jangling relationship, then we might catch from this accident the unease we felt in the *estero*. There is a stretch of coast country below Monterey which affects all sensitive people profoundly, and if they try to describe their feeling they almost invariably do so in musical terms, in the language of symphonic music. And perhaps here the mind and the nerves are true indices of the reality neither segregated nor understood on an intellectual level. Boodin remarks the essential nobility of philosophy and how it has fallen into disrepute. "Somehow," he says, "the laws of thought must be the laws of things if we are going to attempt a science of reality. Thought and things are part of one evolving matrix, and cannot ultimately conflict."[2]

[2] A Realistic Universe, p. xviii. 1931. Macmillan, New York.

And in a unified-field hypothesis, or in life, which is a unified field of reality, everything is an index of everything else. And the truth of mind and the way mind is must be an index of things, the way things are, however much one may stand against the other as an index of the second or irregular order, rather than as a harmonic or first-order index. These two types of indices may be compared to the two types of waves, for indices are symbols as primitive as waves. The first wave-type is the regular or cosine wave, such as tide or undulations of light or sound or other energy, especially where the output is steady and unmixed. These waves may be progressive—increasing or diminishing—or they can seem to be stationary, although deeply some change or progression may be found in all oscillation. All terms of a series must be influenced by the torsion of the first term and by the torsion of the end, or change, or stoppage of the series. Such waves as these may be predictable as the tide is. The second type, the irregular for the while, such as graphs of rainfall in a given region, falls into means which are the functions of the length of time during which observations have been made. These are unpredictable individually; that is, one cannot say that it will rain or not rain tomorrow, but in ten years one can predict a certain amount of rainfall and the season of it. And to this secondary type mind might be close by hinge and "key-in" indices.

We had had many discussions at the galley table and there had been many honest attempts to understand each other's thinking. There are several kinds of reception possible. There is the mind which lies in wait with traps for flaws, so set that it may miss, through not grasping it, a soundness. There is a second which is not reception at all, but blind flight because of laziness, or because some pattern is disturbed by the processes of the discussion. The best reception of all is that which is easy and relaxed, which says in effect, "Let me absorb this thing. Let me try to understand it without private barriers. When I have understood what you are saying, only then will I subject it to my own scrutiny and my own criticism." This is the finest of all critical approaches and the rarest.

The smallest and meanest of all is that which, being frightened or outraged by thinking outside or beyond its pattern, revenges itself senselessly; leaps on a misspelled word or a mispronunciation, drags tricky definition in by the scruff of the neck, and, ranging like a small unpleasant dog, rags and tears the structure to shreds. We have known a critic to base a vicious criticism on a misplaced letter in a word, when actually he was venting rage on an idea he hated. These are the suspicious ones, the self-protective ones, living lives of difficult defense, insuring themselves against folly with folly—stubbornly self-protective at too high a cost.

Ideas are not dangerous unless they find seeding place in some earth more profound than the mind. Leaders and would-be leaders are so afraid that the *idea* "communism" or the *idea* "Fascism" may lead to revolt, when actually they are ineffective without the black earth of discontent to grow in. The strike-raddled businessman may lean toward strikeless Fascism, forgetting that it also eliminates him. The rebel may yearn violently for the freedom from capitalist domination expected in a workers' state, and ignore the fact that such a state is free from rebels. In each case the idea is dangerous only when planted in unease and disquietude. But being so planted, growing in such earth, it ceases to be idea and becomes emotion and then religion. Then, as in most things teleologically approached, the wrong end of the animal is attacked. Lucretius, striking at the teleology of his time, was not so far from us. "I shall untangle by what power the steersman nature guides the sun's courses, and the meanderings of the moon, lest we, percase, should fancy that of own free will they circle their perennial courses round, timing their motions for increase of crops and living creatures, or lest we should think they roll along by any plan of gods. For even *those* men who have learned full well that godheads lead a long life free of care, if yet meanwhile they wonder by what plans things can go on (and chiefly yon high things observed o'erhead on the ethereal coasts), again are hurried back unto the fears of old religion and adopt again harsh masters, deemed almighty,—wretched men, unwitting what can be and what cannot, and by what law to each its scope prescribed, its boundary stone that clings so deep in Time."[3]

In the afternoon we sailed down the coast carefully, for the sandbars were many and some of them uncharted. It was a shallow sea again, and the blueness of deep water had changed to the gray-

[3] Lucretius, *On the Nature of Things,* W. E. Leonard translation, Everyman's Library, 1921, p. 190.

green of sand and shallows. Again we saw manta rays, but not on the surface this day, and the hunt had gone out of us. Tex did not even get out his new harpoon. Perhaps the crew were homesick now. They had seen Guaymas, they were bloated with stories, and they wanted to get back to Monterey to tell them. We would stop at no more towns, see no more people. The inland water of Agiabampo was our last stop, and then quickly home. The shore was low and hot and humid, covered with brush and mangroves. The sea was sterile, or populated with sharks and rays. No algae adhered to the sand bottom, and we were sad in this place after the booming life of the other side. We sailed all afternoon and it was evening when we came to anchorage five miles offshore in the safety of deeper water. We would edge in with the leadline in the morning.

28

APRIL 11

At ten o'clock we moved toward the northern side of the entrance of Agiabampo estuary. The sand-bars were already beginning to show with the lowering tide. Tiny used the leadline on the bow while Sparky was again on the crow's-nest where he could watch for the shallow water. Tony would not approach closer than a mile from the entrance, leaving as always a margin of safety.

When we anchored, five of us got into the little skiff, filling it completely. Any rough water would have swamped us. Sparky and Tiny rowed us in, competing violently with each other, which gave a curious twisting course to the boat.

Agiabampo is a great lagoon with a narrow seaward entrance. There is a little town ten miles in on the northern shore which we did not even try to reach. The entrance is intricate and obstructed with many shoals and sand-bars. It would be difficult without local knowledge to bring in a boat of any draft. We moved in around the northern shore; there were dense thickets of mangrove with little river-like entrances winding away into them. We saw great expanses of sand flat and the first extensive growth of eel-grass we had found.[1] But the eel-grass, which ordinarily shelters a great variety of animal life, was here not very rich at all. We saw the depressions where *botete,* the poison fish, lay. And there were great numbers of sting-rays, which made us walk very carefully, even in rubber boots, for a slash with the tail-thorn of a sting-ray can easily pierce a boot.

The sand banks near the entrance were deeply cut by currents. High in the intertidal many grapsoid crabs[2] lived in slanting burrows about eighteen inches deep. There were a great many of the huge stalk-eyed conchs and the inevitable big hermit crabs living

[1] The true *Zostera marina* according to Dr. Dawson, botanist at the University of California, who remarks that it had not been reported previously so far south.

[2] *Ocypode occidentalis.*

in the cast-off conch shells. Farther in, there were numbers of *Chione* and the blue-clawed swimming crabs. They seemed even cleverer and fiercer here than at other places. Some of the eel-grass was sexually mature, and we took it for identification. On this grass there were clusters of snail eggs, but we saw none of the snails that had laid them. We found one scale-worm,[3] a magnificent specimen in a *Cerianthus*-like tube. There were great numbers of tube-worms in the sand. The wind was light or absent while we collected, and we could see the bottom everywhere. On the exposed sand-bars birds were feeding in multitudes, possibly on the tube-worms. Along the shore, oyster-catchers hunted the burrowing crabs, diving at them as they sat at the entrances of their houses. It was not a difficult collecting station; the pattern, except for the eel-grass, was by now familiar to us although undoubtedly there were many things we did not see. Perhaps our eyes were tired with too much looking.

As soon as the tide began its strong ebb we got into the skiff and started back to the *Western Flyer.* Collecting in narrow-mouthed estuaries, we are always wrong with the currents, for we come in against an ebbing tide and we go out against the flow. It was heavy work to defeat this current. The Sea-Cow gave us a hand and we rowed strenuously to get outside.

That night we intended to run across the Gulf and start for home. It was good to be running at night again, easier to sleep with the engine beating. Tiny at the wheel inveighed against the waste of fish by the Japanese. To him it was a waste complete, a loss of something. We discussed the widening and narrowing picture. To Tiny the fisherman, having as his function not only the catching of fish but the presumption that they would be eaten by humans, the Japanese were wasteful. And in that picture he was very correct. But all the fish actually were eaten; if any small parts were missed by the birds they were taken by the detritus-eaters, the worms and cucumbers. And what they missed was reduced by the bacteria. What was the fisherman's loss was a gain to another group. We tried to say that in the macrocosm nothing is wasted, the equation always balances. The elements which the fish elaborated into an individuated physical organism, a microcosm, go back again into

[3] *Polyodontes oculea.*

the undifferentiated macrocosm which is the great reservoir. There is not, nor can there be, any actual waste, but simply varying forms of energy. To each group, of course, there must be waste—the dead fish to man, the broken pieces to gulls, the bones to some and the scales to others—but to the whole, there is no waste. The great organism, Life, takes it all and uses it all. The large picture is always clear and the smaller can be clear—the picture of eater and eaten. And the large equilibrium of the life of a given animal is postulated on the presence of abundant larvae of just such forms as itself for food. Nothing is wasted; "no star is lost."

And in a sense there is no over-production, since every living thing has its niche, *a posteriori,* and God, in a real, non-mystical sense, sees every sparrow fall and every cell utilized. What is called "over-production" even among us in our manufacture of articles is only over-production in terms of a status quo, but in the history of the organism, it may well be a factor or a function in some great pattern of change or repetition. Perhaps some cells, even intellectual ones, must be sickened before others can be well. And perhaps with us these production climaxes are the therapeutic fevers which cause a rush of curative blood to the sickened part. Our history is as much a product of torsion and stress as it is of unilinear drive. It is amusing that at any given point of time we haven't the slightest idea of what is happening to us. The present wars and ideological changes of nervousness and fighting seem to have direction, but in a hundred years it is more than possible it will be seen that the direction was quite different from the one we supposed. The limitation of the seeing point in time, as well as in space, is a warping lens.

Among men, it seems, historically at any rate, that processes of co-ordination and disintegration follow each other with great regularity, and the index of the co-ordination is the measure of the disintegration which follows. There is no mob like a group of well-drilled soldiers when they have thrown off their discipline. And there is no lostness like that which comes to a man when a perfect and certain pattern has dissolved about him. There is no hater like one who has greatly loved.

We think these historical waves may be plotted and the harmonic curves of human group conduct observed. Perhaps out of such observation a knowledge of the function of war and destruction might emerge. Little enough is known about the function of individual pain and suffering, although from its profound organization it is suspected of being necessary as a survival mechanism. And nothing whatever is known of the group pains of the species, although it is not unreasonable to suppose that they too are somehow functions of the surviving species. It is too bad that against even such investigation we build up a hysterical and sentimental barrier. Why do we so dread to think of our species as a species? Can it be that we are afraid of what we may find? That human self-love would suffer too much and that the image of God might prove to be a mask? This could be only partly true, for if we could cease to wear the image of a kindly, bearded, interstellar dictator, we might find ourselves true images of his kingdom, our eyes the nebulae, and universes in our cells.

The safety-valve of all speculation is: *It might be so.* And as long as that *might* remains, a variable deeply understood, then speculation does not easily become dogma, but remains the fluid creative thing it might be. Thus, a valid painter, letting color and line, observed, sift into his eyes, up the nerve trunks, and mix well with his experience before it flows down his hand to the canvas, has made his painting say, "It might be so." Perhaps his critic, being not so honest and not so wise, will say, "It is not so. The picture is damned." If this critic could say, "It is not so with me, but that might be because my mind and experience are not identical with those of the painter," that critic would be the better critic for it, just as that painter is a better painter for knowing he himself is in the pigment.

We tried always to understand that the reality we observed was partly us; the speculation, our product. And yet if somehow, "The laws of thought must be the laws of things," one can find an index of reality even in insanity.

We sailed a compass course in the night and before daylight a deep fog settled on us. Tony stopped the engine and let us drift, and the dawn came with the thick fog still about us. Tiny and Sparky had the watch, and as the dawn broke, they heard surf and reported it. We came out of our bunks and went up on the deckhouse just as the fog lifted. There was an island half a mile away. Then Tony said, "Did you keep the course I gave you?" Tiny insisted that they had, and Tony said, "If that is so, you have discovered an island, and a big one, because the chart shows no island here." He went on delicately, "I want to congratulate you. We'll call it 'Colletto and Enea Island.' " Tony continued silkily, "But you know God-damn well you didn't keep the course. You know you forgot, and are a good many miles off course." Sparky and Tiny did not argue. They never claimed the island, nor mentioned it again. It developed that it was Espíritu Santo Island, and would have been a prize if they had discovered it, but some Spaniards had done that several hundred years before.

San Gabriel Bay was near us, its coral sand dazzlingly white, and a good reef projecting and a mangrove swamp along part of the coast. We went ashore for this last collecting station. The sand was so white and the water so clear that we took off our clothes and plunged about. The animals here had been affected by the white sand. The crabs were pale and nearly white, and all the animals, even the starfish, were strangely colored. There were stretches of this blinding sand alternating with bouldery reef and mangrove. In the center of the little bay, a fine big patch of green coral almost emerged from the water. It was green and brown coral in great heads, and there were *Phataria* and many club-spined urchins on the heads. There were multitudes of the clam *Chione* just under the surface of the sand, very hard to find until we discovered that every clam had a tiny veil of pale-green algae growing on the front of each valve and sticking up above the sand. Then we took a great number of them.

Near the beds of clams lived heart-urchins with vicious spines.[4] These too were buried in the sand, and to dig for the clams was to be stabbed by the heart urchins, and to be stung badly. There were many hachas here with their clustered colonies of associated fauna. We found solitary and clustered zoanthidean anemones, possibly the same we had been seeing in many variations. We found light-colored *Callinectes* crabs and one of the long snake-like sea-cucumbers[5] such as the ones we had taken at Puerto Escondido. On the rocky reef there were anemones, limpets, and many barnacles. The most common animal on the reef was a membranous tube-worm[6]

[4] *Lovenia cordiformis.*
[5] *Euapta godeffroyi.*
[6] *Megalomma mushaensis.*

with tentacles like a serpulid's. These tentacles were purple and brown, but when approached they were withdrawn and the animal became sand-colored. The mangrove region here was rich. The roots of the trees, impacted with rocks, maintained a fine group of crabs and cucumbers. Two large, hairy grapsoid crabs[7] lived highest in the littoral. They were very fast and active and difficult to catch, and when caught, battled fiercely and ended up by autotomizing.

There was also a *Panopeus*-like crab, *Xanthodius hebes*, but dopey and slow. We found great numbers of porcelain crabs and snapping shrimps. There were barnacles on the reef and on the roots of the mangroves; two new ophiurans and a large sea-hare, besides a miscellany of snails and clams. It was a rich haul, this last day. The sun was hot and the sand pleasant and we were comfortable except for mosquito bites. We played in the water a long time when we were tired of collecting.

When once the engine started now, it would not stop until we reached San Diego. We were reluctant to go back. This balance in time is one of the very few occasions when we have the right of "yes" and "no," and even now the cards were stacked against "yes."

At last we picked up the collecting buckets and the little crowbars and all the tubes, and we rowed slowly back to the *Western Flyer*. Even then, we had difficulty in starting. Someone was overboard swimming in the beautiful water all the time. Tony and Tex, who had been eager to get home, were reluctant now that it was upon them. We had all felt the pattern of the Gulf, and we and the Gulf had established another pattern which was a new thing composed of it and us. At last, and with sorrow, Tex started the engine and the anchor came up for the last time.

All afternoon we stowed and lashed equipment, set the corks in hundreds of glass tubes and wrapped them in paper toweling, screwed tight the caps of jars, tied down the skiffs, and finally dropped the hatch cover in place. We covered the bookcase with triple tarpaulin, and one last time overcame the impulse to throw the Sea-Cow overboard. Then we were under way, sailing southward toward the Cape. The swordfish jumped in the afternoon light, flashing like heliographs in the distance. We took back our old watches that night, and the engine drummed happily and drove

[7] *Geograpsus* and *Goniopsis*.

us through a calm sea. In the morning the tip of the Peninsula was on our right. Behind us the Gulf was sunny and calm, but out in the Pacific a heavy threatening line of clouds hung.

Then a crazy literary thing happened. As we came opposite the Point there was one great clap of thunder, and immediately we hit the great swells of the Pacific and the wind freshened against us. The water took on a gray tone.

29

APRIL 13

At three A.M. Pacific time we passed the light on the false cape and made our new course northward, and the sky was gray and threatening and the wind increased. The Gulf was blotted out for us—the Gulf that was thought and work and sunshine and play. This new world of the Pacific took hold of us and we thought again of an unseen person on the deckhouse, some kind of symbol person—to a sailor, a ghost, a premonition, a feeling in human form.

We could not yet relate the microcosm of the Gulf with the macrocosm of the sea. As we went northward the gray waves rolled up and the *Western Flyer* stubbed her nose into them and the white spray flew over us. The day passed and a new night came and the sea grew more stern. Now we plunged like a nervous horse, and no step could be taken without a steadying hand. The galley was in confusion, for a can of olive oil had leaped from its stand and flooded the floor. On the stove, the coffee pot slipped back and forth between its bars.

Over the surface of the heaving sea the birds flew landward, zigzagging to cover themselves in the wave troughs from the wind. The man at the wheel was the lucky one, for he had a grip against the pitching. He was closest to the boat and to the rising storm. He was the receiver, but also he was the giver and his hand was on the course.

What was the shape and size and color and tone of this little expedition? We slipped into a new frame and grew to be a part of it, related in some subtle way to the reefs and beaches, related to the little animals, to the stirring waters and the warm brackish lagoons. This trip had dimension and tone. It was a thing whose boundaries seeped through itself and beyond into some time and space that was more than all the Gulf and more than all our lives. Our fingers turned over the stones and we saw life that was like our life.

On the deckhouse we held the rails for support, and the blunt nose of the boat fought into the waves and the gray-green water struck us in the face. Some creative thing had happened, a real tempest in our small teapot minds. But boiling water still produces steam, whether in a watch-glass or in a turbine. It is the same stuff—weak and dissipating or explosive, depending on its use. The shape of the trip was an integrated nucleus from which weak strings of thought stretched into every reachable reality, and a reality which reached into us through our perceptive nerve trunks. The laws of thought seemed really one with the laws of things. There was some quality of music here, perhaps not to be communicated, but sounding clear and huge in our minds. The boat plunged and shook herself, and rivers of swirling water ran down the scuppers. Below in the hold, packed in jars, were thousands of little dead animals, but we did not think of them as trophies, as things cut off from the tide pools of the Gulf, but rather as drawings, incomplete and imperfect, of how it had been there. The real picture of how it had been there and how we had been there was in our minds, bright with sun and wet with sea water and blue or burned, and the whole crusted over with exploring thought. Here was no service to science, no naming of unknown animals, but rather—we simply liked it. We liked it very much. The brown Indians and the gardens of the sea, and the beer and the work, they were all one thing and we were that one thing too.

The *Western Flyer* hunched into the great waves toward Cedros Island, the wind blew off the tops of the whitecaps, and the big guy wire, from bow to mast, took up its vibration like the low pipe on a tremendous organ. It sang its deep note into the wind.

APPENDIX: ABOUT ED RICKETTS

Just about dusk one day in April 1948 Ed Ricketts stopped work in the laboratory in Cannery Row. He covered his instruments and put away his papers and filing cards. He rolled down the sleeves of his wool shirt and put on the brown coat which was slightly small for him and frayed at the elbows.

He wanted a steak for dinner and he knew just the market in New Monterey where he could get a fine one, well hung and tender.

He went out into the street that is officially named Ocean View Avenue and is known as Cannery Row. His old car stood at the gutter, a beat-up sedan. The car was tricky and hard to start. He needed a new one but could not afford it at the expense of other things.

Ed tinkered away at the primer until the ancient rusty motor coughed and broke into a bronchial chatter which indicated that it was running. Ed meshed the jagged gears and moved away up the street.

He turned up the hill where the road crosses the Southern Pacific Railways track. It was almost dark, or rather that kind of mixed light and dark which makes it very difficult to see. Just before the crossing the road takes a sharp climb. Ed shifted to second gear, the noisiest gear, to get up the hill. The sound of his motor and gears blotted out every other sound. A corrugated iron warehouse was on his left, obscuring any sight of the right of way.

The Del Monte Express, the evening train from San Francisco, slipped around from behind the warehouse and crashed into the old car. The cow-catcher buckled in the side of the automobile and pushed and ground and mangled it a hundred yards up the track before the train stopped.

Ed was conscious when they got him out of the car and laid him on the grass. A crowd had collected of course—people from the train and more from the little houses that hug the track.

In almost no time a doctor was there. Ed's skull had a crooked look and his eyes were crossed. There was blood around his mouth, and his body was twisted, distorted—wrong, as though seen under an untrue lens.

The doctor got down on one knee and leaned over. The ring of people was silent.

Ed asked, "How bad is it?"

"I don't know," the doctor said. "How do you feel?"

"I don't feel much of anything," Ed said.

Because the doctor knew him and knew what kind of a man he was, he said, "That's shock, of course."

"Of course!" Ed said, and his eyes began to glaze.

They edged him onto a stretcher and took him to the hospital. Section hands pried his old car off the cow-catcher and pushed it aside, and the Del Monte Express moved slowly into the station at Pacific Grove, which is the end of the line.

Several doctors had come in and more were phoning, wanting to help because they all loved him. The doctors knew it was very serious, so they gave him ether and opened him up to see how bad it was. When they finished they knew it was hopeless. Ed was all messed up—spleen broken, ribs shattered, lungs punctured, concussion of the skull. It might have been better to let him go out under the ether, but the doctors could not give up, any more than could the people gathered in the waiting room of the hospital. Men who knew better began talking about miracles and how anything could happen. They reminded each other of cases of people who had got well when there was no reason to suppose they could. The surgeons cleaned Ed's insides as well as possible and closed him up. Every now and then one of the doctors would go out to the waiting room, and it was like facing a jury. There were lots of people out there, sitting waiting, and their eyes all held a stone question.

The doctors said things like, "Doing as well as can be expected" and "We won't be able to tell for some time but he seems to be making progress." They talked more than was necessary, and the people sitting there didn't talk at all. They just stared, trying to get adjusted.

The switchboard was loaded with calls from people who wanted to give blood.

The next morning Ed was conscious but very tired and groggy

from ether and morphine. His eyes were washed out and he spoke with great difficulty. But he did repeat his first question.

"How bad is it?"

The doctor who was in the room caught himself just as he was going to say some soothing nonsense, remembering that Ed was his friend and that Ed loved true things and knew a lot of true things too, so the doctor said, "Very bad."

Ed didn't ask again. He hung on for a couple of days because his vitality was very great. In fact he hung on so long that some of the doctors began to believe the things they had said about miracles when they knew such a chance to be nonsense. They noted a stronger heartbeat. They saw improved color in his cheeks below the bandages. Ed hung on so long that some people from the waiting room dared to go home to get some sleep.

And then, as happens so often with men of large vitality, the energy and the color and the pulse and the breathing went away silently and quickly, and he died.

By that time the shock in Monterey had turned to dullness. He was dead and had to be got rid of. People wanted to get rid of him quickly and with dignity so they could think about him and restore him again.

On a small rise not far from the Great Tide Pool near Lighthouse Point there is a small chapel and crematory. Ed's closed coffin was put in that chapel for part of an afternoon.

Naturally no one wanted flowers, but the greatest fear was that someone might say a speech or make a remark about him—good or bad. Luckily it was all over so quickly that the people who ordinarily make speeches were caught unprepared.

A large number of people drifted into the chapel, looked for a few moments at the coffin, and then walked away. No one wanted company. Everyone wanted to be alone. Some went to the beach by the Great Tide Pool and sat in the coarse sand and blindly watched the incoming tide creeping around the rocks and tumbling in over the seaweed.

A kind of anesthesia settled on the people who knew Ed Ricketts. There was not sorrow really but rather puzzled questions—what are we going to do? how can we rearrange our lives now? Everyone who knew him turned inward. It was a strange thing—quiet and strange. We were lost and could not find ourselves.

It is going to be difficult to write down the things about Ed Ricketts that must be written, hard to separate entities. And anyone who knew him would find it difficult. Maybe some of the events are imagined. And perhaps some very small happenings may have grown out of all proportion in the mind. And then there is the personal impact. I am sure that many people, seeing this account, will be sure to say, "Why, that's not true. That's not the way he was at all. He was this way and this." And the speaker may go on to describe a person this writer did not know at all. But no one who knew him will deny the force and influence of Ed Ricketts. Everyone near him was influenced by him, deeply and permanently. Some he taught how to think, others how to see or hear. Children on the beach he taught how to look for and find beautiful animals in worlds they had not suspected were there at all. He taught everyone without seeming to.

Nearly everyone who knew him has tried to define him. Such things were said of him as, "He was half-Christ and half-goat." He was a great teacher and a great lecher—an immortal who loved women. Surely he was an original and his character was unique, but in such a way that everyone was related to him, one in this way and another in some different way. He was gentle but capable of ferocity, small and slight but strong as an ox, loyal and yet untrustworthy, generous but gave little and received much. His thinking was as paradoxical as his life. He thought in mystical terms and hated and distrusted mysticism. He was an individualist who studied colonial animals with satisfaction.

We have all tried to define Ed Ricketts with little success. Perhaps it would be better to put down the mass of material from our memories, anecdotes, quotations, events. Of course some of the things will cancel others, but that is the way he was. The essence lies somewhere. There must be some way of finding it.

Finally there is another reason to put Ed Ricketts down on paper. He will not die. He haunts the people who knew him. He is always present even in the moments when we feel his loss the most.

One night soon after his death a number of us were drinking beer in the laboratory. We laughed and told stories about Ed, and suddenly one of us said in pain, "We'll have to let him go! We'll have to release him and let him go." And that was true not for Ed but for ourselves. We can't keep him, and still he will not go away.

Maybe if I write down everything I can remember about him, that will lay the ghost. It is worth trying anyway. It will have to be true or it can't work. It must be no celebration of his virtues, because, as was said of another man, he had the faults of his virtues. There can be no formula. The simplest and best way will be just to remember—as much as I can.

The statistics on Ed Ricketts would read: Born in Chicago, played in the streets, went to public school, studied biology at the University of Chicago. Opened a small commercial laboratory in Pacific Grove, California. Moved to Cannery Row in Monterey. Degrees—Bachelor of Science only; clubs, none; honors, none. Army service—both World Wars. Killed by a train at the age of fifty-two. Within that frame he went a long way and burned a deep scar.

I was sitting in a dentist's waiting room in New Monterey, hoping the dentist had died. I had a badly aching tooth and not enough money to have a good job done on it. My main hope was that the dentist could stop the ache without charging too much and without finding too many other things wrong.

The door to the slaughterhouse opened and a slight man with a beard came out. I didn't look at him closely because of what he held in his hand, a bloody molar with a surprisingly large piece of jawbone sticking to it. He was cursing gently as he came through the door. He held the reeking relic out to me and said, "Look at that god-damned thing." I was already looking at it. "That came out of me," he said.

"Seems to be more jaw than tooth," I said.

"He got impatient, I guess. I'm Ed Ricketts."

"I'm John Steinbeck. Does it hurt?"

"Not much. I've heard of you."

"I've heard of you, too. Let's have a drink."

That was the first time I ever saw him. I had heard that there was an interesting man in town who ran a commercial laboratory, had a library of good music, and interests wider than invertebratology. I had wanted to come across him for some time.

We did not think of ourselves as poor then. We simply had no money. Our food was fairly plentiful, what with fishing and planning and a minimum of theft. Entertainment had to be improvised without benefit of currency. Our pleasures consisted in conversation, walks, games, and parties with people of our own financial nonexistence. A real party was dressed with a gallon of thirty-nine-cent wine, and we could have a hell of a time on that. We did not know any rich people, and for that reason we did not like them and were proud and glad we didn't live *that* way.

We had been timid about meeting Ed Ricketts because he was rich people by our standards. This meant that he could depend on a hundred to a hundred and fifty dollars a month and he had an automobile. To us this was fancy, and we didn't see how anyone could go through that kind of money. But we learned.

Knowing Ed Ricketts was instant. After the first moment I knew him, and for the next eighteen years I knew him better than I knew anyone, and perhaps I did not know him at all. Maybe it was that way with all of his friends. He was different from anyone and yet so like that everyone found himself in Ed, and that might be one of the reasons his death had such an impact. It wasn't Ed who had died but a large and important part of oneself.

When I first knew him, his laboratory was an old house in Cannery Row which he had bought and transformed to his purposes. The entrance was a kind of showroom with mounted marine specimens in glass jars on shelves around the walls. Next to this room was a small office, where for some reason the rattlesnakes were kept in cages between the safe and the filing cabinets. The top of the safe was piled high with stationery and filing cards. Ed loved paper and cards. He never ordered small amounts but huge supplies of it.

On the side of the building toward the ocean were two more rooms, one with cages for white rats—hundreds of white rats, and reproducing furiously. This room used to get pretty smelly if it was not cleaned with great regularity—which it never was. The other rear room was set up with microscopes and slides and the equipment for making and mounting and baking the delicate microorganisms which were so much a part of the laboratory income. In the basement there was a big stockroom with jars and tanks for preserving the larger animals, and also the equipment for embalming and injecting the cats, dogfish, frogs, and other animals that were used by dissection classes.

This little house was called Pacific Biological Laboratories, Inc., as strange an operation as ever outraged the corporate laws of California. When, after Ed's death, the corporation had to be liquidated, it was impossible to find out who owned the stock, how much of it there was, or what it was worth. Ed kept the most careful collecting notes on record, but sometimes he would not open a business letter for weeks.

How the business ran for twenty years no one knows, but it did run even though it staggered a little sometimes. At times it would spurt ahead with system and efficiency and then wearily collapse for several months. Orders would pile up on the desk. Once during a weary period someone sent Ed a cheesecake by parcel post. He thought it was preserved material of some kind, and when he finally opened it three months later we could not have identified it had it not been that a note was enclosed which said, "Eat this cheesecake at once. It's very delicate."

Often the desk was piled so high with unopened letters that they slid tiredly to the floor. Ed believed completely in the theory that a letter unanswered for a week usually requires no answer, but he went even farther. A letter unopened for a month does not require opening.

Every time some definite statement like that above is set down I think of exceptions. Ed carried on a large and varied correspondence with a number of people. He answered letters quickly and at length, using a typewriter with elite type to save space. The purchase of a typewriter was a long process with him, for much of the type had to be changed from business signs to biologic signs, and he also liked to have some foreign-language signs on his typewriter, tilde for Spanish, accents and cedilla for French, umlaut for German. He rarely used them but he liked to have them.

The days of the laboratory can be split into two periods. The era before the fire and that afterwards. The fire was interesting in many respects.

One night something went wrong with the electric current on the whole water front. Where 220 volts were expected and prepared for, something like two thousand volts suddenly came through. Since in the subsequent suits the electric company was found blameless by the courts, this must be set down to an act of God. What happened was that a large part of Cannery Row burst into flames in a moment. By the time Ed awakened, the laboratory was a sheet of fire. He grabbed his typewriter, rushed to the basement, and got his car out just in time, and just before the building was about ready to crash into its own basement. He had no pants but he had transportation and printing. He always admired his choice. The scientific library, accumulated with such patience and some of it irreplaceable, was gone. All the fine equipment, the microscopes, the museum jars, the stock—everything was gone. Besides typewriter and automobile, only one thing was saved.

Ed had a remarkably fine safe. It was so good that he worried for fear some misguided and romantic burglar might think there was something of value in it and, trying to open it, might abuse and injure its beautiful mechanism. Consequently he not only never locked the safe but contrived a wood block so that it could not be locked. Also, he pasted a note above the combination, assuring all persons that the safe was not locked. Then it developed that there was nothing to put in the safe anyway. Thus the safe became the repository of foods which might attract the flies of Cannery Row, and there were clouds of them drawn to the refuse of the fish canneries but willing to come to other foods. And it must be said that no fly was ever able to negotiate the safe.

But to get back to the fire. After the ashes had cooled, there was the safe lying on its side in the basement where it had fallen when the floor above gave way. It must have been an excellent safe, for when we opened it we found half a pineapple pie, a quarter of a pound of Gorgonzola cheese, and an open can of sardines—all of them except the sardines in good condition. The sardines were a little dry. Ed admired that safe and used to refer to it with affection. He would say that if there *had* been valuable things in the safe it would surely have protected them. "Think how delicate Gorgonzola is," he said. "It couldn't have been very hot inside that safe. The cheese is still delicious."

In spite of a great erudition, or perhaps because of it, Ed had some naive qualities. After the fire there were a number of suits against the electric company, based on the theory, later proved wrong, that if the fires were caused by error or negligence on the part of the company, the company should pay for the damage.

Pacific Biological Laboratories, Inc., was one of the plaintiffs in this suit. Ed went over to Superior Court in Salinas to testify. He told the truth as clearly and as fully as he could. He loved true things and believed in them. Then he became fascinated by the trial and the jury and he spent much time in court, inspecting the legal system with the same objective care he would have lavished on a new species of marine animal.

Afterwards he said calmly and with a certain wonder, "You see how easy it is to be completely wrong about a simple matter. It was always my conviction—or better, my impression—that the legal system was designed to arrive at the truth in matters of human and property relationships. You see, I had forgotten or never considered one thing. Each side wants to win, and that factor warps any original intent to the extent that the objective truth of the matter disappears in emphasis. Now you take the case of this fire," he went on. "Both sides wanted to win, and neither had any interest in, indeed both sides seemed to have a kind of abhorrence for, the truth." It was an amazing discovery to him and one that required thinking out. Because he loved true things, he thought everyone did. The fact that it was otherwise did not sadden him. It simply interested him. And he set about rebuilding his laboratory and replacing his books with an antlike methodicalness.

Ed's use of words was unorthodox and, until you knew him, somewhat startling. Once, in getting a catalogue ready, he wanted to advise the trade that he had plenty of hagfish available. Now the hagfish is a most disgusting animal both in appearance and texture, and some of its habits are nauseating. It is a perfect animal horror. But Ed did not feel this, because the hagfish has certain functions which he found fascinating. In his catalogue he wrote, "Available in some quantities, delightful and beautiful hagfish."

He admired worms of all kinds and found them so desirable that, searching around for a pet name for a girl he loved, he called her "Wormy." She was a little huffy until she realized that he was using not the adjective but a diminutive of the noun. His use of this word meant that he found her pretty, interesting, and desirable. But still it always sounded to the girl like an adjective.

Ed loved food, and many of the words he used were eating words. I have heard him refer to a girl, a marine animal, and a plain song as "delicious."

His mind had no horizons. He was interested in everything. And there were very few things he did not like. Perhaps it would be well to set down the things he did not like. Maybe they would be some kind of key to his personality, although it is my conviction that there is no such key.

Chief among his hatreds was old age. He hated it in other people and did not even conceive of it in himself. He hated old women and would not stay in a room with them. He said he could smell them. He had a remarkable sense of smell. He could smell a mouse in a room, and I have seen him locate a rattlesnake in the brush by smell.

He hated women with thin lips. "If the lips are thin—where will there be any fullness?" he would say. His observation was certainly physical and open to verification, and he seemed to believe in its accuracy and so do I, but with less vehemence.

He loved women too much to take any nonsense from the thin-lipped ones. But if a girl with thin lips painted on fuller ones with lipstick, he was satisfied. "Her intentions are correct," he said. "There is a psychic fullness, and sometimes that can be very fine."

He hated hot soup and would pour cold water into the most beautifully prepared bisque.

He unequivocally hated to get his head wet. Collecting animals in the tide pools, he would be soaked by the waves to his eyebrows, but his head was invariably covered and safe. In the shower he wore an oilskin sou'wester—a ridiculous sight.

He hated one professor whom he referred to as "old jingle ballicks." It never developed why he hated "old jingle ballicks."

He hated pain inflicted without good reason. Driving through the streets one night, he saw a man beating a red setter with a rake handle. Ed stopped the car and attacked the man with a monkey wrench and would have killed him if the man had not run away.

Although slight in build, when he was angry Ed had no fear and could be really dangerous. On an occasion one of our cops was pistol-whipping a drunk in the middle of the night. Ed attacked the cop with his bare hands, and his fury was so great that the cop released the drunk.

This hatred was only for reasonless cruelty. When the infliction of pain was necessary, he had little feeling about it. Once during the depression we found we could buy a live sheep for three dollars. This may seem incredible now but it was so. It was a great deal of food and even for those days a great bargain. Then we had the sheep and none of us could kill it. But Ed cut its throat with no emotion whatever, and even explained to the rest of us who were upset that bleeding to death is quite painless if there is no fear involved. The pain of opening a vein is slight if the instrument is sharp, and he had opened the jugular with a scalpel and had not frightened the animal, so that our secondary or empathic pain was probably much greater than that of the sheep.

His feeling for psychic pain in normal people also was philosophic. He would say that nearly everything that can happen to people not only does happen but has happened for a million years. "Therefore," he would say, "for everything that can happen there is a channel or mechanism in the human to take care of it—a channel worn down in prehistory and transmitted in the genes."

He disliked time intensely unless it was part of an observation or an experiment. He was invariably and consciously late for appointments. He said he had once worked for a railroad where his whole life had been regulated by a second hand and that he had then conceived his disgust, a disgust for exactness in time. To my knowledge, that is the only time he ever spoke of the railroad experience. If you asked him to dinner at seven, he might get there at nine. On the other hand, if a good low collecting tide was at 6:53, he would be in the tide pool at 6:52.

The farther I get into this the more apparent it becomes to me that no rule was final. He himself was not conscious of any rules of behavior in himself, although he observed behavior patterns in other people with delight.

For many years he wore a beard, not large, and slightly pointed, which accentuated his half-goat, half-Christ appearance. He had started wearing the beard because some girl he wanted thought he had a weak chin. He didn't have a weak chin, but as long as she thought so he cultivated his beard. This was probably during the period of the prognathous Arrow Collar men in the advertising pages. Many girls later he was still wearing the beard because he was used to it. He kept it until the Army made him shave it off in the Second World War. His beard sometimes caused a disturbance. Small boys often followed Ed, baaing like sheep. He developed a perfect defense against this. He would turn and baa back at them, which invariably so embarrassed the boys that they slipped shyly away.

Ed had a strange and courteous relationship with dogs, although he never owned one or wanted to. Passing a dog on the street, he greeted it with dignity and, when driving, often tipped his hat and smiled and waved at dogs on the sidewalk. And damned if they didn't smile back at him. Cats, on the other hand, did not arouse any enthusiasm in him. However, he always remembered one cat with admiration. It was in the old days before the fire when Ed's father was still alive and doing odd jobs about the laboratory. The cat in question took a dislike to Ed's father and developed a spite tactic which charmed Ed. The cat would climb up on a shelf and pee on Ed's father when he went by—the cat did it not once but many times.

Ed regarded his father with affection. "He has one quality of genius," Ed would say. "He is always wrong. If a man makes a million decisions and judgments at random, it is perhaps mathematically tenable to suppose that he will be right half the time and wrong half the time. But you take my father—he is wrong all of the time about everything. That is a matter not of luck but of selection. That requires genius."

Ed's father was a rather silent, shy, but genial man who took so many aspirins for headaches that he had developed a chronic acetanilide poisoning and the quaint dullness that goes with it. For many years he worked in the basement stockroom, packing specimens to be shipped and even mounting some of the larger and less delicate forms. His chief pride, however, was a human fetus which he had mounted in a museum jar. It was to have been the lone child of a Negress and a Chinese. When the mother succumbed to a lover's quarrel and a large dose of arsenic administered by person or persons unknown, the autopsy revealed her secret, and her secret was acquired by Pacific Biological. It was much too far advanced to be of much value for study so Ed's father inherited it. He crossed its little legs in a Buddha pose, arranged its hands in an attitude of semi-prayer, and fastened it securely upright in the museum jar. It was rather a startling figure, for while it had negroid features, the preservative had turned it to a pale ivory color. It was Dad Ricketts' great pride. Children and many adults made pilgrimages to the basement to see it. It became famous in Cannery Row.

One day an Italian woman blundered into the basement. Although she did not speak any English, Dad Ricketts naturally thought she had come to see his prize. He showed it to her; whereupon, to his amazement and embarrassment, she instantly undressed to show him her fine scar from a Caesarian section.

Cats were a not inconsiderable source of income to Pacific Biological Laboratories, Inc. They were chloroformed, the blood drained, and embalming fluid and color mass injected in the venous and arterial systems. These finished cats were sold to schools for study of anatomy.

When an order came in for, say, twenty-five cats, there was only one way to get them, since the ASPCA will not allow the raising of cats for laboratory purposes. Ed would circulate the word among the small boys of the neighborhood that twenty-five cents apiece would be paid for cats. It saddened Ed a little to see how venially warped the cat-loving small boys of Monterey were. They sold their own cats, their aunts' cats, their neighbors' cats. For a few days there would be scurrying footsteps and soft thumps as cats in gunny sacks were secretly deposited in the basement. Then guileless and innocent-faced little catacides would collect their quarters and rush for Wing Chong's grocery for pop and cap pistols. No matter what happened, Wing Chong made some small profit.

Once a lady who liked cats very much, if they were the better sort of cats, remarked to Ed, "Of course I realize that these things are necessary. I am very broad-minded. But, thank heaven, you do not get pedigreed cats."

Ed reassured her by saying, "Madam, that's about the only kind I do get. Alley cats are too quick and intelligent. I get the sluggish stupid cats of the rich and indulgent. You can look through the basement and see whether I have yours—yet." That friendship based on broad-mindedness did not flourish.

If there were a complaint and a recognition Ed always gave the cat back. Once two small boys who had obviously read about the oldest cheat in the world worked it twice on Ed before he realized it. One of them sold the cat and collected, the other came in crying and got the cat back. They should have got another cat the third time. If they had been clever and patient they would have made a fortune, but even Ed recognized a bright yellow cat with a broken tail the third time he bought it.

Everyone, so Ed said, has at least one biologic theory, and some people develop many. Ed was very tolerant of these flights of theoretic fancy. A strange group flowed through the laboratory.

There were, for instance, the people who suddenly discovered parallels in nature, like the man who conceived the thought that tuna, which is called commercially "Chicken of the Sea," might be related to chickens, because, as he said, "Their eyes look alike." Ed's reply to this man was that he rarely liked to make a positive statement, but in this case he was willing to venture the conviction that there was no very close relation between chickens and tunas.

One day there came into the laboratory a young Chinese, dressed in the double-breasted height of fashion, smelling of lily of the valley, and bringing a mysterious air with him. He was about twenty-three and his speech was that of an American high-school boy. He suggested darkly that he would like to see Ed alone. Ed happily joined the mystery and indicated that I was his associate and the sharer of his secrets. We found ourselves speaking in heavy, pregnant whispers.

Our visitor asked, "Have you got any cat blood?"

"No, not right now," Ed replied. "It is true I do draw off the blood when I inject cats. What do you want it for?"

Our visitor said tightly, "I'm making an experiment." Then, to prove that we could trust his judgment and experience, he flipped his lapel to show the badge of a detective correspondence school. And he drew out his diploma to back up the badge. We were delighted with him. But he would not explain what he needed the cat blood for. Ed promised that he would save some blood from the next series of cats. We all nodded mysteriously back and forth and our visitor left quietly, walking on his toes.

Mysteries were constant at the laboratory. A thing happened one night which I later used as a short story. I wrote it just as it happened. I don't know what it means and do not even answer the letters asking what its philosophic intent is. It just happened. Very briefly, this is the incident. A woman came in one night wanting to buy a male rattlesnake. It happened that we had one and knew it was a male because it had recently copulated with another snake in the cage. The woman paid for the snake and then insisted that it be fed. She paid for a white rat to be given it. Ed put the rat in the cage. The snake struck and killed it and then unhinged its jaws preparatory to swallowing it. The frightening thing was that the woman, who had watched the process closely, moved her jaws and stretched her mouth just as the snake was doing. After the rat was swallowed, she paid for a year's supply of rats and said she would come back. But she never did come back. What happened or why

I have no idea. Whether the woman was driven by a sexual, a religious, a zoophilic, or a gustatory impulse we never could figure. When I wrote the story just as it happened there were curious reactions. One librarian wrote that it was not only a bad story but the worst story she had ever read. A number of orders came in for snakes. I was denounced by a religious group for having a perverted imagination, and one man found symbolism of Moses smiting the rock in the account.

I shall mention only a few other of the mysteries. There was the persecution with flowers, for example. Someone who must have been watching the laboratory waited until we were out on several occasions and then placed a line of white flowers across the doorstep. This happened a number of times and seems to have been meant as a hex. Such a curse is practiced by some northern Indians to bring death to anyone who steps over the flowers. But who put them there and whether that was the intention we never found out.

During the time when the Klan was spreading its sheets all over the nation the laboratory got its share of attention. Small red cards with the printed words, "We are watching *you*, K.K.K.," were slipped under the door on several occasions.

Mysteries had a bad effect on Ed Ricketts. He hated all thoughts and manifestations of mysticism with an intensity which argued a basic and undefeatable belief in them. He refused to have his fortune told or his palm read even in fun. The play with a Ouija board drove him into a nervous rage. Ghost stories made him so angry that he would leave a room where one was being told.

In the course of time Ed's father died. There was an intercom phone between the basement and the upstairs office. Once after his father's death Ed admitted to me that he had a waking nightmare that the intercom phone would ring, that he would lift the receiver and hear his father's voice on the other end. He had dreamed of this, and it was becoming an obsession with him. I suggested that someone might play a practical joke and that it might be a good idea to disconnect the phone. This he did instantly, but he went further and removed both phones. "It would be worse disconnected," he said. "I couldn't stand that."

I think that if anyone had played such a joke, Ed would have been very ill from shock. The white flowers bothered him a great deal.

I have said that his mind had no horizons, but that is untrue.

He forbade his mind to think of metaphysical or extra-physical matters, and his mind refused to obey him.

Life on Cannery Row was curious and dear and outrageous. Across the street from Pacific Biological was Monterey's largest, most genteel and respected whorehouse. It was owned and operated by a very great woman who was beloved and trusted by all who came in contact with her except those few whose judgment was twisted by a limited virtue. She was a large-hearted woman and a law-abiding citizen in every way except one—she did violate the nebulous laws against prostitution. But since the police didn't seem to care, she felt all right about it and even made little presents in various directions.

During the depression Madam paid the grocery bills for most of the destitute families on Cannery Row. When the Chamber of Commerce collected money for any cause and businessmen were assessed at ten dollars, Madam was always nicked for a hundred. The same was true for any mendicant charity. She halfway paid for the widows and orphans of policemen and firemen. She was expected to and did contribute ten times the ordinary amount toward any civic brainstorm of citizens who pretended she did not exist. Also, she was a wise and tolerant pushover for any hard-luck story. Everyone put the bee on her. Even when she knew it was a fake she dug down.

Ed Ricketts maintained relations of respect and friendliness with Madam. He did not patronize the house. His sex life was far too complicated for that. But Madam brought many of her problems to him, and he gave her the best of his thinking and his knowledge, both scientific and profane.

There seems to be a tendency toward hysteria among girls in such a house. I do not know whether hysterically inclined types enter the business or whether the business produces hysteria. But often Madam would send a girl over to the laboratory to talk to Ed. He would listen with great care and concern to her troubles, which were rarely complicated, and then he would talk soothingly to her and play some of his favorite music to her on his phonograph. The girl usually went back reinforced with his strength. He never moralized in any way. He would be more likely to examine the problem carefully, with calm and clarity, and to lift the horrors out of it by easy examination. Suddenly the girl would discover that she was not alone, that many other people had the same problems—in a word that her misery was not unique. And then she usually felt better about it.

There was a tacit but strong affection between Ed and Madam. She did not have a license to sell liquor to be taken out. Quite often Ed would run out of beer so late at night that everything except Madam's house was closed. There followed a ritual which was thoroughly enjoyed by both parties. Ed would cross the street and ask Madam to sell him some beer. She invariably refused, explaining every time that she did not have a license. Ed would shrug his shoulders, apologize for asking, and go back to the lab. Ten minutes later there would be soft footsteps on the stairs and a little thump in front of the door and then running slippered steps down again. Ed would wait a decent interval and then go to the door. And on his doorstep, in a paper bag, would be six bottles of ice-cold beer. He would never mention it to Madam. That would have been breaking the rules of the game. But he repaid her with hours of his time when she needed his help. And his help was not inconsiderable.

Sometimes, as happens even in the soundest whorehouse, there would be a fight on a Saturday night—one of those things which are likely to occur when love and wine come together. It was only sensible that Madam would not want to bother the police or a doctor with her little problem. Then her good friend Ed would patch up cut faces and torn ears and split mouths. He was a good operator and there were never any complaints. And naturally no one ever mentioned the matter since he was not a doctor of medicine and had no license to practice anything except philanthropy. Madam and Ed had the greatest respect for each other. "She's one hell of a woman," he said. "I wish good people could be as good."

Just as Madam was the target for every tired heist, so Ed was the fall guy for any illicit scheme that could be concocted by the hustling instincts of some of the inhabitants of Cannery Row. The people of the Row really loved Ed, but this affection did not forbid them from subjecting him to any outrageous scheming that occurred to them. In nearly all cases he knew the game before the play had even started and his hand would be in his pocket before the intricate gambit had come to a request. But he would cautiously wait out the pitch before he brought out the money. "It gives them so much pleasure to earn it," he would say.

He never gave much. He never *had* much. But in spite of his wide experience in chicanery, now and then he would be startled into admiration by some particularly audacious or imaginative approach to the problem of a touch.

One evening while he was injecting small dogfish in the basement, one of his well-known clients came to him with a face of joy.

"I am a happy man," the hustler proclaimed, and went on to explain how he had arrived at the true philosophy of rest and pleasure.

"You think I've got nothing, Eddie," the man lectured him. "But you don't know from my simple outsides what I've got inside."

Ed moved restlessly, waiting for the trap.

"I've got peace of mind, Eddie. I've got a place to sleep, not a palace but comfortable. I'm not hungry very often. And best of all I've got friends. I guess I'm gladdest of all for my friends."

Ed braced himself. Here it comes, he thought.

"Why, Ed," the client continued, "some nights I just lay in my bed and thank God for my blessings. What does a man need, Eddie—a few things like food and shelter and a few little tiny vices, like liquor and women—and tobacco—"

Ed could feel it moving in on him. "No liquor," he said.

"I ain't drinking," the client said with dignity, "didn't you hear?"

"How much?" Ed asked.

"Only a dime, Eddie boy. I need a couple of sacks of tobacco. I don't mind using the brown papers on the sacks. I *like* the brown papers."

Ed gave him a quarter. He was delighted. "Where else in the world could you find a man who would lavish care and thought and art and emotion on a lousy dime?" he said. He felt that it had been worth more than a quarter, but he did not tell his client so.

On another occasion Ed was on his way across the street to Wing Chong's grocery for a couple of quarts of beer. Another of his clients was sitting comfortably in the gutter in front of the store. He glanced casually at the empty quart bottles Ed carried in his hand.

"Say Doc," he said, "I'm having a little trouble peeing. What's a good diuretic?"

Ed fell into that hole. "I never needed to think beyond beer," he said.

The man looked at the bottles in Ed's hand and raised his shoulders in a gesture of helplessness. And only then did Ed realize that he had been had. "Oh, come on in," he said, and he bought beer for both of them.

Afterward he said admiringly, "Can you imagine the trouble he went to for that beer? He had to look up the word diuretic, and then he had to plan to be there just when I went over for beer. And he had to read my mind quite a bit. If any part of his plan failed, it all failed. I think it is remarkable."

The only part of it that was not remarkable was planning to be there when Ed went for beer. He went for beer pretty often. Sometimes when he overbought and the beer got warm, he took it back and Wing Chong exchanged it for cold beer.

The various hustlers who lived by their wits and some work in the canneries when they had time were an amazing crew. Ed never got over his admiration for them.

"They have worked out my personality and my resistances to a fine mathematical point," he would say. "They know me better than I know myself, and I am not uncomplicated. Over and over, their analysis of my possible reaction is accurate."

He was usually delighted when one of these minor triumphs took place. It never cost him much. He always tried to figure out in advance what the attack on his pocket would be. At least he always knew the end. Every now and then the audacity and freedom of thought and invention of his loving enemy would leave him with a sense of wonder.

Now and then he hired some of the boys to collect animals for him and paid them a fixed price, so much for frogs, so much for snakes or cats.

One of his collectors we will call Al. That was not his name. An early experience with Al gave Ed a liking for his inventiveness. Ed needed cats and needed them quickly. And Al got them and got them quickly—all fine mature cats and, only at the end of the operation did Ed discover, all tomcats. For a long time Al held out his method but finally he divulged it in secret. Since Al has long since gone to his maker and will need no more cats, his secret can be told.

"I made a double trap," he said, "a little cage inside a big cage. Then in the little cage I put a nice lady cat in a loving condition. And, Eddie, sometimes I'd catch as many as ten tomcats in one night. Why, hell, Ed, that exact same kind of trap catches me every Saturday night. That's where I got the idea."

Al was such a good collector that after a while he began to do odd jobs around the lab. Ed taught him to inject dogfish and to work the ball mill for mixing color mass and to preserve some of the less delicate animals. Al became inordinately proud of his work and began to use a mispronounced scientific vocabulary and put on a professorial air that delighted Ed. He got to trusting Al although he knew Al's persistent alcoholic history.

Once when a large number of dogfish came in Ed left them for Al to inject while he went to a party. It was a late party. Ed returned to find all the lights on in the basement. The place was a wreck. Broken glass littered the floor, a barrel of formaldehyde was tipped over and spilled, museum jars were stripped from the shelves and broken. A whirlwind had gone through. Al was not there but Al's pants were, and also an automobile seat which was never explained.

In a white fury Ed began to sweep up the broken glass. He was well along when Al entered, wearing a long overcoat and a pair of high rubber boots. Ed's rage was terrible. He advanced on Al.

"You son-of-a-bitch!" he cried. "I should think you could stay sober until you finished work!"

Al held up his hand with senatorial dignity. "You go right ahead, Eddie," he said. "You call me anything you want, and I forgive you."

"Forgive me?" Ed screamed. He was near to murder.

Al silenced him with a sad and superior gesture. "I deserve it, Eddie," he said. "Go ahead—call me lots of names. I only regret that they will not hurt my feelings."

"What in hell are you talking about?" Ed demanded uneasily.

Al turned and parted the tails of his overcoat. He was completely naked except for the rubber boots.

"Eddie boy," he said, "I have been out calling socially in this condition. Now, Eddie, if I could do that, I must be pretty insensitive. Nothing you can call me is likely to get under my thick skin. And I forgive you."

Ed's anger disappeared in pure wonder. And afterward he said, "If that Al had turned the pure genius of his unique mind to fields other than cadging drinks, there is no limit to what he might have done." And then he continued, "But no. He has chosen a difficult and crowded field and he is a success in it. Any other career, international banking for instance, might have been too easy for Al."

Al was married, but his wife and family did not exercise a restraining influence on him. His wife finally used the expedient of putting Al in jail when he was on one of his beauties.

Al said one time, "When they hire a new cop in Monterey they give him a test. They send him down Cannery Row, and if he can't pick me up he don't get the job."

Al detested the old red stone Salinas jail. It was gloomy and unsanitary, he said. But then the county built a beautiful new jail, and the first time Al made sixty days he was gone seventy-five. He came back to Monterey enthusiastic.

"Eddie," he said, "they got radios in the cells. And that new sheriff's a pushover at euchre. When my time was up the sheriff owed me eighty-six bucks. I couldn't run out on the game. A sheriff can make it tough on a man. It took me fifteen days to lose it back so it wouldn't look too obvious. But you can't win from a sheriff, Eddie—not if you expect to go back."

Al went back often until his wife finally tumbled to the fact that Al preferred jail to home life. She visited Ed for advice. She was a red-eyed, unkempt little woman with a runny nose.

"I work hard and try to make ends meet," she said bitterly. "And all the time Al's over in Salinas taking his ease in the new jail. I can't let him go to jail any more. He likes it." She was all frayed from having Al's children and supporting them.

For once Ed had no answer. "I don't know what you can do," he said. "I'm stumped. You could kill him—but then you wouldn't have any fun any more."

A complicated social structure existed on Cannery Row. One had to know or there were likely to be errors in procedure and protocol. You could not speak to one of the girls from Madam's if you met her on the street. You might have talked to her all night, but it was bad manners to greet her outside.

From the windows of the laboratory Ed and I watched a piece of social cruelty which has never been bettered in Scarsdale. Across the street in the lot between the whorehouse and Wing Chong's grocery, there were a number of rusty pipes, a boiler or two, and some great timbers, all thrown there by the canneries. A number of the free company of Cannery Row slept in the big pipes, and when the sun was warm they would come out to sit like lizards on the timbers. There they held social commerce. They borrowed dimes back and forth, shared tobacco, and if anyone brought a pint of liquor into sight, it meant that he not only wanted to share it but intended to. They were a fairly ragged set of men, their clothing of blue denim almost white at knees and buttocks from pure erosion. They were, as Ed said, the Lotus Eaters of our era, successful in their resistance against the nervousness and angers and frustrations of our time.

Ed regarded these men with the admiration he had for any animal, family, or species that was successful in survival and happiness factors.

We had many discussions about these men. Ed held that one couldn't tell from a quick look how successful a species is.

"Consider now," he would say, "if you look superficially, you would say that the local banker or the owner of a cannery or even the mayor of Monterey is the successful and surviving individual. But consider their ulcers, consider the heart trouble, the blood pressure in that group. And then consider the bums over there—cirrhosis of the liver I will grant will have its toll, but not the other things." He would cluck his tongue in admiration. "It is a rule in paleontology," he would say, "that over-armor, and/or over-ornamentation are symptoms of extinction in a species. You have only to consider the great reptiles, the mammoth, etc. Now those bums have no armor and practically no ornament, except here and there a pair of red and yellow sleeve garters. In our whole time pattern those men may be the ones who will deliver our species from the enemies within and without which attack it."

But much as he liked the bums, he was grieved at their social cruelty toward George, the pimp of the whorehouse.

George was well built, a snappy dresser, and very polite. He had complete extra-legal police powers over the girls in the house and an arguable access to any or all of them. He might even treat a friend. He had dark wavy hair, a good salary, he ate in the house, and he clipped several of the girls for their money. In other words, he was rich. He was a good bouncer with an enviable reputation for in-fighting, and—when the problem grew more confused—a triumphant record of eye-gouging, booting, and kneeing. In a word, one would have thought him a happy man—one would, unless one knew the true soul of George, as we came to.

George was lonely. He wanted the company of men, the camaraderie and warmth and roughness and good feeling and arguments of men. He got very tired of a woman's world of perfumes and periods, of hysterics and noisy mysteries and permanents. Perhaps he had no one to boast to of his superiority over women, and it bothered him.

We watched him try to associate with the bums sitting on the timber in the sun, and they would have none of him. They considered a pimp as abysmally beneath them socially. When George wandered up through the weeds and sat with the boys they would turn away from him. They did not insult him or tell him to go away, but they would not associate with him. If an argument was going on when he arrived, it would stop and a painful silence would take its place.

George recognized his ostracism and he was sad and hangdog about it. We, watching from the window, could see it in his wilting posture and his fawning gestures. We could hear it in his too loud laughter at a mildly amusing joke. Ed shook his head over this injustice. He had hoped for better from the boys.

"I don't know why I thought they would be better," he said. "Of course, being bums does give them advantages, but why should I expect them to be above all smallness just because they are bums? I guess it was just a romantic hopefulness." And he said, "I knew a man who believed all whores were honest just because they were whores. Time and again he got rolled—once a girl even stole his clothes, but he would not give up his conviction. It had become an article of faith, and you can't give such a thing up because it is yourself. I must re-examine my feeling about the boys," he said.

We watched George fall back, in his craven loneliness, on bribery. He bought whisky and passed it around. He loaned money like a crazy man. The bums accepted George's bribes but they would not accept George.

Ed Ricketts did not ordinarily meddle in the affairs of his neighbors but he brooded about George.

One afternoon he confronted the boys on the timber. "Why don't you be nice to him?" he said. "He's a lonely man. He wants to be friends with you. You are putting a mark on him that may warp and sour his whole life. He won't be any good to anyone. I wouldn't be surprised if you were responsible for his death."

To which Whitey No. 2 (there were two Whiteys, known as Whitey No. 1 and Whitey No. 2) replied, "Now, Doc, you're not asking us to associate with a pimp, are you? Nobody likes a pimp."

It must be noted that when the hustlers spoke to Ed formally he was Doc. When they hustled him, he was Ed, Eddie, or Eddie boy.

I don't think that Ed had any idea how accurate his prediction was. But not very long after this George killed himself with an ice pick in the kitchen of the whorehouse. And when Ed berated the boys for having been one of the causes of his death, Whitey No. 1 echoed Whitey No. 2's words.

"Hell, we can't help it, Doc. You just can't be friendly with a pimp."

Ed mused sadly, "I find it rather hard to believe that the boys were moved by any moral consideration. It must have been an unscalable social barrier that no argument could overleap." And he said, "White chicks will kill a black chick every time. But I do hope it isn't as simple as that."

Ed's association with Wing Chong, the Chinese grocer, and later, after Wing Chong's death, with his son, was one of mutual respect. Ed could always get credit and for long periods of time. And sometimes he needed it. Once we tried to compute how many gallons of beer had crossed the street in the years of our association, but we soon gave up as the figures mounted. We didn't even want to know.

Ed had many friends, and in addition he attracted some people from the lunatic fringe, like the Chinese detective and the snake woman. There were others who used him as a source of information.

One afternoon the phone rang and a woman's voice asked, "Dr.

Ricketts, can you tell me the name of a tropical fish with so many spines on the dorsal fin and so many on the ventral? The name begins with an L."

"Not offhand," said Ed, "but I'll be glad to look it up for you if you want to call back in half an hour." He went to work, saying, "Lovely voice—fine throaty voice."

Twenty minutes later the phone rang again and the fine throaty voice said, "Dr. Ricketts, never mind. I worked it out from the horizontals."

He never did meet the puzzle-worker with the throaty voice.

In appearance and temperament Ed was a remarkably unmilitary man, but in spite of this he was drafted for service in both World Wars. One would have thought that his complete individuality and his uniqueness of approach to all problems would have caused him to go crazy in the organized mediocrity of the Army. Actually the exact opposite was true. He was a successful soldier. In spite of itself, the Army—at least that part of it which sheltered him—was gradually warped in his favor and for his comfort. He was quite happy in the Army in both wars.

He described his military experience in the first World War to me with satisfaction. "I was young then," he said, "and I am amazed that I showed such good sense. I have often thought," he went on, "that if any big company like General Motors or Standard Oil should start a private army, no public army would stand a chance against it. A private company is organized to do something or to produce something, profit or gold or steel. It has a direction. But a public army is made up of millions of individuals all working for themselves. Some want promotions, some want to steal, some want personal power or glory, and some want simply to get out. Very few have any interest in winning a war."

He told me about his first war experience. "I gave it a good deal of thought before I decided what to be," he said. "As I said, I was young then, but I have always admired my choice. Literacy was not terribly high in 1917, and it was comparatively easy for me to become company clerk without any danger of being driven into officer's training school. I definitely did not want to be an officer. No one wanted the job of company clerk.

"People are singularly blind," he continued. "It escaped the greed and self-interest of the other men that the company clerk makes out the passes and that if the captain and lieutenants happen to have hobbies like golf or women, this duty and even the selections are left in the hands of an efficient company clerk." He sighed with pleasure. He had enjoyed the Army. "In almost no time," he said, "the rumor got about that I liked whisky. It became quite common knowledge. And do you know, when I was demobilized I had over three hundred pints left, and that in a time of prohibition, if you will remember."

A little venom crept into his voice. "You know," he said in an outraged tone, "there was one christing son-of-a-bitch who complained to the captain about me. Can you imagine that? He put it on a moral basis. He didn't drink. I wonder why nondrinkers are so often vicious."

"What happened?" I asked.

"He was a silly man," Ed said. "He didn't get a single pass for eighteen months. He wrote complaint after complaint. He was a very silly man."

"But how about the complaints?"

"If he had given it any thought he would have realized that complaints go through the hands of the company clerk." He chuckled. "I guess I should not bear a grudge," he said, "but I still don't like that man. Word got about—you know how rumors move in the Army. Anyway, the word got out that the good, kind company clerk was being persecuted. I guess the poor fellow had a rough time of it—from latrines to kitchen police to the brig. I think it ruined his whole military career. I'm pretty sure it ruined his stomach. A very silly man."

I have always felt that drafting Ed in the Second World War was spiteful on the part of the draft board. He was one week under forty-six when his call came, and his birthday had passed when he was examined. I think there were people in Monterey who were jealous of him. He was really not good soldier material from any point of view. He wore a beard, which is frowned on by Army psychiatrists. The doctor who examined him came from the interview puzzled and worried, but he passed Ed, and the Army made him shave off his beard.

He did not resent being drafted because he remembered the first war with such pleasure.

"I thought that with my subsequent experience and maturity I might be all right," he said.

Because of his long laboratory experience they put him in charge of the venereal disease section of the induction center at Monterey. This job had its compensations. He could go home every night and he had complete charge of an inexhaustible medicine chest. He was still in no danger of being hustled off to officer's training school. Ed didn't want to command men. He wanted to associate with them. His commanding officer had a hobby—whether golf or women I do not know, but it was strong enough so that he let Ed do all the work.

Ed liked that and did a very good job with his section. Possibly because of the medicine chest a little group of passionate admirers clung to him and protected him and defended him against any possible charge that Ed didn't get to work before ten in the morning and sometimes went away for long weekends.

Quite early in his second hitch in the Army Ed got tired of the sameness of laboratory alcohol and grapefruit juice. With his unlimited medicine chest, he began to experiment. Now another rumor crept about the Presidio of Monterey that a fabulous drink had been invented. It had a strange effect. No one had tasted or felt anything quite like it. It was called "Ricketts' Folly." It was said that the commanding officer of the unit, and he a major at that, after two drinks of it had marched smartly and with no hint of stagger right into a wall, and that he had made a short heroic speech as he slid to the ground.

After Ed was safely and honorably discharged I asked him about the drink that had achieved a notoriety as far east as Chicago and that was discussed with hushed respect on the beachheads of the Pacific.

"Well, actually it was very simple," he said. "It's components were not complicated and it *was* delicious. I never could figure why it had such a curious and sometimes humorous effect. It was nothing but alcohol, codeine, and grenadine. It was a pretty drink too. You know," he said, "it made every other kind of liquor seem kind of weak and flabby."

This account of Ed Ricketts goes seesawing back and forth chronologically and in every other way. I did not intend when I started to departmentalize him, but now that seems to be a good method. He was so complex and many-faceted that perhaps the best method will be to go from one facet of him to another so that from all the bits a whole picture may build itself for me as well as for others.

Ed had more fun than nearly anyone I have ever known, and he had deep sorrows also, which will be treated later. As long as we are on the subject of drinking I will complete that department.

Ed loved to drink, and he loved to drink just about anything. I don't think I ever saw him in the state called drunkenness, but twice he told me he had no memory of getting home to the laboratory at all. And even on those nights one would have had to know him well to be aware that he was affected at all. Evidences of drinking were subtle. He smiled a little more broadly. His voice became a little higher in pitch, and he would dance a few steps on tiptoe, a curious pigeon-footed mouse step. He liked every drink that contained alcohol and, except for coffee which he often laced with whisky, he disliked every drink that did not contain alcohol. He once estimated that it had been twelve years since he had tasted water without some benign addition.

At one time when bad teeth and a troublesome love affair were running concurrently, he got a series of stomach-aches which were diagnosed as a developing ulcer. The doctor put him on a milk diet and ordered him off all alcohol. A sullen sadness fell on the laboratory. It was a horrid time. For a few days Ed was in a state of dismayed shock. Then his anger rose at the cruelty of a fate that could do this to him. He merely disliked and distrusted water, but he had an active and fierce hatred for milk. He found the color unpleasant and the taste ugly. He detested its connotations.

For a few days he forced a little milk into his stomach, complaining bitterly the while, and then he went back to see the doctor. He explained his dislike for the taste of milk, giving as its basis some pre-memory shock amounting to a trauma. He thought this dislike for milk might have driven him into the field of marine biology since no marine animals but whales and their family of sea cows give milk and he had never had the least interest in any of the Cetaceans. He said that he was afraid the cure for his stomach-aches was worse than the disease and finally he asked if it would be all right to add a few drops of aged rum to the milk just to kill its ugly taste. The doctor perhaps knew he was fighting a losing battle. He gave in on the few drops of rum.

We watched the cure with fascination as day by day the ratio changed until at the end of a month Ed was adding a few drops of milk to the rum. But his stomach-aches had disappeared. He never liked milk, but after this he always spoke of it with admiration as a specific for ulcers.

There were great parties at the laboratory, some of which went on for days. There would come a time in our poverty when we needed a party. Then we would gather together the spare pennies. It didn't take very many of them. There was a wine sold in Monterey for thirty-nine cents a gallon. It was not a delicate-tasting wine and sometimes curious things were found in the sludge on the bottom of the jug, but it was adequate. It added a gaiety to a party and it never killed anyone. If four couples got together and each brought a gallon, the party could go on for some time and toward the end of it Ed would be smiling and doing his tippy-toe mouse dance.

Later, when we were not so poor, we drank beer or, as Ed preferred it, a sip of whisky and a gulp of beer. The flavors, he said, complemented each other.

Once on my birthday there was a party at the laboratory that lasted four days. We really needed a party. It was fairly large, and no one went to bed except for romantic purposes. Early in the morning at the end of the fourth day a benign exhaustion had settled on the happy group. We spoke in whispers because our vocal chords had long since been burned out in song.

Ed carefully placed half a quart of beer on the floor beside his bed and sank back for a nap. In a moment he was asleep. He had consumed perhaps five gallons since the beginning of the party. He slept for about twenty minutes, then stirred, and without opening his eyes groped with his hand for the beer bottle. He found it, sat up, and took a deep drink of it. He smiled sweetly and waved two fingers in the air in a kind of benediction.

"There's nothing like that first taste of beer," he said.

Not only did Ed love liquor. He went further. He had a deep suspicion of anyone who did not. If a non-drinker shut up and minded his own business and did not make an issue of his failing, Ed could be kind to him. But alas, a laissez-faire attitude is very uncommon in teetotalers. The moment one began to spread his poison Ed experienced a searing flame of scorn and rage. He believed that anyone who did not like to drink was either sick and/

or crazy or had in him some obscure viciousness. He believed that the soul of a non-drinker was dried up and shrunken, that the virtuous pose of the non-drinker was a cover for some nameless and disgusting practice.

He had somewhat the same feeling for those who did not or pretended they did not love sex, but this field will be explored later.

If pressed, Ed would name you the great men, great minds, great hearts and imaginations in the history of the world, and he could not discover one of them who was a teetotaler. He would even try to recall one single man or woman of much ability who did not drink and like liquor, and he could never light on a single name. In all such discussions the name of Shaw was offered, and in answer Ed would simply laugh, but in his laughter there would be no admiration for that abstemious old gentleman.

Ed's interest in music was passionate and profound. He thought of it as deeply akin to creative mathematics. His taste in music was not strange but very logical. He loved the chants of the Gregorian mode and the whole library of the plain song with their angelic intricacies. He loved the masses of William Byrd and Palestrina. He listened raptly to Buxtehude, and he once told me that he thought the *Art of the Fugue* of Bach might be the greatest of all music up to our time. Always "up to our time." He never considered anything finished or completed but always continuing, one thing growing on and out of another. It is probable that his critical method was the outgrowth of his biologic training and observation.

He loved the secular passion of Monteverde, and the sharpness of Scarlatti. His was a very broad appreciation and a curiosity that dug for music as he dug for his delicious worms in a mud flat. He listened to music with his mouth open as though he wanted to receive the tones even in his throat. His forefinger moved secretly at his side in rhythm.

He could not sing, could not carry a tune or reproduce a true note with his voice, but he could hear true notes. It was a matter of sorrow to him that he could not sing.

Once we bought sets of tuning forks and set them in rubber to try to reteach ourselves the forgotten mathematical scale. And Ed's ear was very aware in recognition although he could not make his voice come even near to imitating the pitch. I never heard him whistle. I wonder whether he could. He would try to hum melodies, stumbling over the notes, and then he would smile helplessly when his ear told him how badly he was doing it.

He thought of music as something incomparably concrete and dear. Once, when I had suffered an overwhelming emotional upset, I went to the laboratory to stay with him. I was dull and speechless with shock and pain. He used music on me like medicine. Late in the night when he should have been asleep, he played music for me on his great phonograph—even when I was asleep he played it, knowing that its soothing would get into my dark confusion. He played the curing and reassuring plain songs, remote and cool and separate, and then gradually he played the sure patterns of Bach, until I was ready for more personal thought and feeling again, until I could bear to come back to myself. And when that time came, he gave me Mozart. I think it was as careful and loving medication as has ever been administered.

Ed's reading was very broad. Of course he read greatly in his own field of marine invertebratology. But he read hugely otherwise. I do not know where he found the time. I can judge his liking only by the things he went back to—translations of Li Po and Tu Fu, that greatest of all love poetry, the *Black Marigolds*—and *Faust,* most of all *Faust.* Just as he thought the *Art of the Fugue* might be the greatest music up to our time, he considered *Faust* the greatest writing that had been done. He enlarged his scientific German so that he could read Faust and hear the sounds of the words as they were written and taste their meanings. Ed's mind seems to me to have been a timeless mind, not modern and not ancient. He loved to read Layamon aloud and Beowulf, making the words sound as fresh as though they had been written yesterday.

He had no religion in the sense of creed or dogma. In fact he distrusted all formal religions, suspecting them of having been fouled with economics and power and politics. He did not believe in any God as recognized by any group or cult. Probably his God could have been expressed by the mathematical symbol for an expanding universe. Surely he did not believe in an after life in any sense other than chemical. He was suspicious of promises of an after life, believing them to be sops to our fear or hope artificially supplied.

Economics and politics he observed with the same interested

detachment he applied to the ecological relationships and balances in a tide pool.

For a time after the Russian Revolution he watched the Soviet with the pleased interest of a terrier seeing its first frog. He thought there might be some new thing in Russia, some human progression that might be like a mutation in the nature of the species. But when the Revolution was accomplished and the experiments ceased and the Soviets steadied and moved inexorably toward power and the perpetuation of power through applied ignorance and dogmatic control of the creative human spirit, he lost interest in the whole thing. Now and then he would take a sampling to verify his conclusions as to the direction. His last hope for that system vanished when he wrote to various Russian biologists, asking for information from their exploration of the faunal distribution on the Arctic Sea. He then discovered that they not only did not answer, they did not even get his letters. He felt that any restriction or control of knowledge or conclusion was a dreadful sin, a violation of first principles. He lost his interest in Marxian dialectics when he could not verify in observable nature. He watched with a kind of amused contempt while the adepts warped the world to fit their pattern. And when he read the conclusions of Lysenko, he simply laughed without comment.

Very many conclusions Ed and I worked out together through endless discussion and reading and observation and experiment. We worked together, and so closely that I do not now know in some cases who started which line of speculation since the end thought was the product of both minds. I do not know whose thought it was.

We had a game which we playfully called speculative metaphysics. It was a sport consisting of lopping off a piece of observed reality and letting it move up through the speculative process like a tree growing tall and bushy. We observed with pleasure how the branches of thought grew away from the trunk of external reality. We believed, as we must, that the laws of thought parallel the laws of things. In our game there was no stricture of rightness. It was an enjoyable exercise on the instruments of our minds, improvisations and variations on a theme, and it gave the same delight and interest that discovered music does. No one can say, "This music is the only music," nor would we say, "This thought is the only thought," but rather, "This is *a* thought, perhaps well or ill formed, but *a* thought which is a real thing in nature."

Once a theme was established we subjected observable nature to it. The following is an example of our game—one developed quite a long time ago.

We thought that perhaps our species thrives best and most creatively in a state of semi-anarchy, governed by loose rules and half-practiced mores. To this we added the premise that over-integration in human groups might parallel the law in paleontology that over-armor or over-ornamentation are symptoms of decay and disappearance. Indeed, we thought, over-integration *might be* the symptom of human decay. We thought: there is no creative unit in the human save the individual working alone. In pure creativeness, in art, in music, in mathematics, there are no true collaborations. The creative principle is a lonely and an individual matter. Groups can correlate, investigate, and build, but we could not think of any group that has ever created or invented anything. Indeed, the first impulse of the group seems to be to destroy the creation and the creator. But integration, or the designed group, seems to be highly vulnerable.

Now with this structure of speculation we would slip examples on the squares of the speculative graphing paper.

Consider, we would say, the Third Reich or the Politburo-controlled Soviet. The sudden removal of twenty-five key men from either system could cripple it so thoroughly that it would take a long time to recover, if it ever could. To preserve itself in safety such a system must destroy or remove all opposition as a danger to itself. But opposition is creative and restriction is non-creative. The force that feeds growth is therefore cut off. Now, the tendency to integration must constantly increase. And this process of integration must destroy all tendencies toward improvisation, must destroy the habit of creation, since this is sand in the bearings of the system. The system then must, if our speculation is accurate, grind to a slow and heavy stop. Thought and art must be forced to disappear and a weighty traditionalism take its place. Thus we would play with thinking. A too greatly integrated system or society is in danger of destruction since the removal of one unit may cripple the whole.

Consider the blundering anarchic system of the United States, the stupidity of some of its lawmakers, the violent reaction, the slowness of its ability to change. Twenty-five key men destroyed could make the Soviet Union stagger, but we could lose our congress, our president, and our general staff and nothing much would have happened. We would go right on. In fact we might be better for it.

That is an example of the game we played. Always our thinking was prefaced with, "It might be so!" Often a whole night would draw down to a moment while we pursued the fireflies of our thinking.

Ed spoke sometimes of a period he valued in his life. It was after he had left home and entered the University of Chicago. He had not liked his home life very well. The rules that he had known were silly from his early childhood were finally removed.

"Adults, in their dealing with children, are insane," he said. "And children know it too. Adults lay down rules they would not think of following, speak truths they do not believe. And yet they expect children to obey the rules, believe the truths, and admire and respect their parents for this nonsense. Children must be very wise and secret to tolerate adults at all. And the greatest nonsense of all that adults expect children to believe is that people learn by experience. No greater lie was ever revered. And its falseness is immediately discerned by children since their parents obviously have not learned anything by experience. Far from learning, adults simply become set in a maze of prejudices and dreams and sets of rules whose origins they do not know and would not dare inspect for fear the whole structure might topple over on them. I think children instinctively know this," Ed said. "Intelligent children learn to conceal their knowledge and keep free of this howling mania."

When he left home, he was free at last, and he remembered his first freedom with a kind of glory. His freedom was not one of idleness.

"I don't know when I slept," he said. "I don't think there was time to sleep. I tended furnaces in the early morning. Then I went to class. I had lab all afternoon, then tended furnaces in the early evening. I had a job in a little store in the evening and got some studying done then, until midnight. Well, then I was in love with a girl whose husband worked nights, and naturally I didn't sleep much from midnight until morning. Then I got up and tended furnaces and went to class. What a time," he said, "what a fine time that was."

It is necessary in any kind of picture of Ed Ricketts to give some account of his sex life since that was by far his greatest drive. His life was saturated with sex and he was to a very great extent preoccupied with it. He gave it a monumental amount of thought and time and analysis. It will be no violation to discuss this part of his life since he had absolutely no shyness about discussing it himself.

To begin with, he was a hyper-thyroid. His metabolic rate was abnormally high. He had to eat at very frequent intervals or his body revolted with pain and anger. He was, during the time I knew him, and, I gather, from the very beginning, as concupiscent as a bull terrier. His sexual output and preoccupation was or purported to be prodigious. I do not know beyond doubt about the actual output. That is hearsay but well authenticated; but certainly his preoccupation with sexual matters was very great.

As far as women were concerned, he was completely without what is generally called "honor." It was not that he was dishonorable. The word simply had no meaning for him if it implied abstemiousness. Any man who left a wife in his care and expected him not to try for her was just a fool. He was compelled to try. The woman might reject him, and he would not be unreasonably importunate, but certainly he would not fail for lack of trying.

When I first met him he was engaged in a scholarly and persistent way in the process of deflowering a young girl. This was a long and careful affair. He not only was interested in a sexual sense, but he had also an active interest in the psychic and physical structure of virginity. There was, I believe, none of the usual sense of triumph at overcoming or being first. Ed's physical basis was a pair of very hot pants, but his secondary motive was an active and highly intellectual interest in the state of virginity and the change involved in abandoning that state. His knowledge of anatomy was large, but, as he was wont to say, the variation in structure is delightfully large, even leaving out abnormalities, and this variation gives a constant interest and surprise to a function which is basically pleasant anyway.

The resistance of this particular virgin was surprising. He did not know whether it was based on some block, or on the old-fashioned reluctance of a normal girl toward defloration, or, as he thought possible, on a distaste for himself personally. He inspected each of these possibilities with patient care. And since he had no shyness about himself, it did not occur to him to have any reluctance about discussing his project with his friends and acquaintances. It is perhaps a fortunate thing that this particular virgin did not hear the discussions. They might have embarrassed her, a matter that did not occur to Ed. Many years later, when she heard about the whole thing, she was of the opinion that she might still be a virgin if she had heard herself so intimately discussed. But by then, it was fortunately, she agreed, far too late.

One thing is certain. Ed did not like his sex uncomplicated. If a girl were unattached and without problems as well as willing, his interest was not large. But if she had a husband or seven children or a difficulty with the law or some whimsical neuroticism in the field of love, Ed was charmed and instantly active. If he could have found a woman who was not only married, but a mother, in prison, and one of Siamese twins, he would have been delighted.

It will be impossible to put down much anecdote concerning his activities. The more interesting affairs were discussed with such freedom, not only by Ed but by any number of amateur referees, that they acquired a certain local fame. This may be perfectly acceptable as confirmed gossip, but in print the protagonists might be inclined to consider the histories libelous, and they surely are.

His taste in women was catholic as long as there were complications and no thin lips. Complexion, color of hair and eyes, shape or size, seemed to make no difference to him. He was singularly open to suggestion.

Ordinarily Ed was able to view his fellow humans with the clear sight of objectivity only slightly warped by like or dislike. He could give the best and most valuable advice based on great knowledge and understanding. However, when the strong winds of love shook him, all this was changed. Then his objectivity was likely to blow sky-high.

The object of his affection herself contributed very little to his picture of her. She was only the physical frame on which he draped a woman. She was like those large faceless dolls on which clothes are made. He built his own woman on this form, created her from the ground up, invented her appearance and built her mind, furnished her with talents and sensitivenesses which were not only astonishing but downright untrue. Then the woman in process was likely to come with surprise to the conclusion that she loved poetry she had never heard of, and could not understand if she had, that she breathed shallowly over music the existence of which was equally unknown to her. She became beautiful but not necessarily in any way that was familiar to her. And her thoughts—these would be likely to surprise her most of all, since she might not have been aware that she had any thoughts at all.

I cannot think of this tendency of Ed's as self-delusion. He simply manufactured the woman he wanted, rather like that enlightened knight in the Welsh tale who made a wife entirely out of flowers. Sometimes the building process went on for quite a long time, and when it was completed everyone—even Ed—was quite confused. But at other times the force of his structure changed the raw material until the girl actually became what he thought her to be. I remember one very sharp example of this.

One of our friends was a sardine fisherman who had an interesting and profitable avocation. The sardine season continues only part of the year, leaving some months of idleness, which is the financial downfall of most fishermen. Our friend, however, was never idle and never broke. He managed, booked, protected, disciplined, and robbed a string of women—never many, rarely over five. He was successful and happy in his hobby. He was our friend and we saw a good deal of him.

This story is to illustrate the force and reasonableness of Ed's woman-building. I don't know how he got confused in this matter but he did, and the subsequent history bears out my belief in his success.

Our friend, in a moment of playfulness, brought one of his clients to a party, a small but unfragile blonde of endurance and experience. Ed met her and in a lapse of reason made an error about her. He went to considerable effort to get her away from her protector. He had an idea that she was not only inexperienced but quite shy—this last probably because she barely had acquired the power of speech and did not trust it as a means of communication. Ed thought her beautiful and young and virginal. He took her away on a vacation. He rebuilt her in his mind. And he tried to seduce her, he tried manfully, persuasively, philosophically, to seduce her.

But he had built too well. In some way he had convinced her that she was what he had mistaken her for and she resisted his advances with maidenly fiber and consistency. At the end of a month he had to give up. He never did get to bed with her. But he took up so much of her time that she had to work very hard to make a stake for our friend when the sardine season was over.

In Ed's ecstasy he was able to make true things which lacked a certain scientific verification. One of his loves, one of his greatest, lasted a number of years. Every night he wrote a letter to his love, sometimes three lines, sometimes ten pages of his small, careful typing. He told me that she did the same, that she wrote to him every day, and he believed it. And I know beyond any doubt that in five years he received from her not more than eight childish scribbled notes. And he truly believed that she wrote to him every day.

Ed's scientific notebooks were very interesting. Among his collecting notes and zoological observations there would be the most outspoken and indelicate observation from another kind of collecting. After his death I had to go through these notebooks before turning them over to Hopkins Marine Station, a branch of Stanford University, as Ed's will directed. I was sorry I had to remove a number, a great number, of the entries from the notebooks. I did not do this because they lacked interest, but it occurred to me that a student delving into Ed's notes for information on invertebratology could emerge with blackmail material on half the female population of Monterey. Ed simply had no reticence about such things. I removed the notes but did not destroy them. They have an interest, I think, above the personalities mentioned. In some future time the women involved may lovingly remember the incidents.

In the back of his car Ed carried an ancient blanket that once had been red but that had faded to a salmon pink from use and exposure. It was a battle-scarred old blanket, veteran of many spreadings on hill and beach. Grass seeds and bits of seaweed were pounded and absorbed into the wool itself. I do not think Ed would have started his car in the evening without his blanket in the back seat.

Before love struck and roiled his vision like a stirred pool, Ed had a fine and appraising eye for a woman. He would note with enthusiasm a well-lipped mouth, a swelling breast, a firm yet cushioned bottom, but he also inquired into other subtleties—the padded thumb, shape of foot, length and structure of finger and toe, plump-lobed ear and angle of teeth, thigh and set of hip and movement in walking too. He regarded these things with joy and thanksgiving. He always was pleased that love and women were what they were or what he imagined them.

But for all of Ed's pleasures and honesties there was a transcendent sadness in his love—something he missed or wanted, a searching that sometimes approached panic. I don't know what it was he wanted that was never there, but I know he always looked for it and never found it. He sought for it and listened for it and looked for it and smelled for it in love. I think he found some of it in music. It was like a deep and endless nostalgia—a thirst and passion for "going home."

He was walled off a little, so that he worked at his philosophy of "breaking through," of coming out through the back of the mirror into some kind of reality which would make the day world dreamlike. This thought obsessed him. He found the symbols of "breaking through" in *Faust*, in Gregorian music, and in the sad, drunken poetry of Li Po. Of the *Art of the Fugue* he would say, "Bach nearly made it. Hear now how close he comes, and hear his anger when he cannot. Every time I hear it I believe that this time he will come crashing through into the light. And he never does—not quite."

And of course it was he himself who wanted so desperately to break through into the light.

We worked and thought together very closely for a number of years so that I grew to depend on his knowledge and on his patience in research. And then I went away to another part of the country but it didn't make any difference. Once a week or once a month would come a fine long letter so much in the style of his speech that I could hear his voice over the neat page full of small elite type. It was as though I hadn't been away at all. And sometimes now when the postman comes I look before I think for that small type on an envelope.

Ed was deeply pleased with the little voyage which is described in the latter part of this book, and he was pleased with the manner of setting it down. Often he would read it to remember a mood or a joke.

His scientific interest was essentially ecological and holistic. His mind always tried to enlarge the smallest picture. I remember his saying, "You know, at first view you would think the rattlesnake and the kangaroo rat were the greatest of enemies since the snake hunts and feeds on the rat. But in a larger sense they must be the best of friends. The rat feeds the snake and the snake selects out the slow and weak and generally thins the rat people so that both species can survive. It is quite possible that neither species could exist without the other." He was pleased with commensal animals, particularly with groups of several species contributing to the survival of all. He seemed as pleased with such things as though they had been created for him.

With any new food or animal he looked, felt, smelled, and tasted. Once in a tide pool we were discussing the interesting fact that nudibranchs, although beautiful and brightly colored and tasty-looking and soft and unweaponed, are never eaten by other animals which should have found them irresistible. He reached under water and picked up a lovely orange-colored nudibranch and put it in his mouth. And instantly he made a horrible face and spat and retched, but he had found out why fishes let these living tidbits completely alone.

On another occasion he tasted a species of free-swimming anemone and got his tongue so badly stung by its nettle cells that he could hardly close his mouth for twenty-four hours. But he would have done the same thing the next day if he had wanted to know.

Although small and rather slight, Ed was capable of prodigies of strength and endurance. He could drive for many hours to arrive at a good collecting ground for a favorable low tide, then work like a fury turning over rocks while the tide was out, then drive back to preserve his catch. He could carry heavy burdens over soft and unstable sand with no show of weariness. He had enormous resistance. It took a train to kill him. I think nothing less could have done it.

His sense of smell was very highly developed. He smelled all food before he ate it, not only the whole dish but each forkful. He invariably smelled each animal as he took it from the tide pool. He spoke of the smells of different animals, and some moods and even thoughts had characteristic odors to him—undoubtedly conditioned by some experience good or bad. He referred often to the smells of people, how individual each one was, and how it was subject to change. He delighted in his sense of smell in love.

With his delicate olfactory equipment, one would have thought that he would be disgusted by so-called ugly odors, but this was not true. He could pick over decayed tissue or lean close to the fetid viscera of a cat with no repulsion. I have seen him literally crawl into the carcass of a basking shark to take its liver in the dark of its own body so that no light might touch it. And this is as horrid an odor as I know.

Ed loved fine tools and instruments, and conversely he had a bitter dislike for bad ones. Often he spoke with contempt of "consumer goods"—things made to catch the eye, to delight the first impression with paint and polish, things made to sell rather than to use. On the other hand, the honest workmanship of a good microscope gave him great pleasure. Once I brought him from Sweden a set of the finest scalpels, surgical scissors, and delicate forceps. I remember his joy in them.

His laboratory practice was immaculate and his living quarters were not clean. It was his custom to say that most people paid too much for things they didn't really want, paid too much in effort and time and thought. "If a swept floor gives you enough pleasure and reward to pay for sweeping it, then sweep it," he said. "But if you do not see it dirty or clean, then it is paying too much to sweep it."

I think he set down his whole code or procedure once in a time of stress. He found himself quite poor and with three children to take care of. In a very scholarly manner, he told the children how they must proceed.

"We must remember three things," he said to them. "I will tell them to you in the order of their importance. Number one and first in importance, we must have as much fun as we can with what we have. Number two, we must eat as well as we can, because if we don't we won't have the health and strength to have as much fun as we might. And number three and third and last in importance, we must keep the house reasonably in order, wash the dishes, and such things. But we will not let the last interfere with the other two."

Ed's feeling for clothes was interesting. He wore Bass moccasins, buckskin-colored and quite expensive. He loved thick soft wool socks and wool shirts that would scratch the hell out of anyone else. But outside of those he had no interest. His clothing was fairly ragged, particularly at elbows and knees. He had one necktie hanging in his closet, a wrinkled old devil of a yellow tint, but no one ever saw him wear it. His clothes he just came by, and the coats were not likely to fit him at all. He was not in the least embarrassed by his clothes. He went everywhere in the same costume. And always he seemed strangely neat. Such was his sense of inner security that he did not seem ill dressed. Often people around him appeared over-dressed. The only time he ever wore a hat was when there was some chance of getting his head wet, and then it was likely to be an oilskin sou'wester. But whatever else he wore or did not wear, there was invariably pinned to his shirt pocket a twenty-power Bausch and Lomb magnifying glass on a little roller chain. He used the glass constantly. It was a very close part of him—one of his techniques of seeing.

Always the paradox is there. He loved nice things and did not care about them. He loved to bathe and yet when the water heater in the laboratory broke down he bathed in cold water for over a year before he got around to having it fixed. I finally mended his leaking toilet tank with a piece of chewing gum which I imagine is still there. A broken window was stuffed with newspaper for several years and never was repaired.

He liked comfort and the chairs in the lab were stiff and miserable. His bed was a redwood box laced with hemp rope on which a thin mattress was thrown. And this bed was not big enough for two. Ladies complained bitterly about his bed, which was not only narrow and uncomfortable but gave out shrieks of protest at the slightest movement.

I used the laboratory and Ed himself in a book called *Cannery Row*. I took it to him in typescript to see whether he would resent it and to offer to make any changes he would suggest. He read it through carefully, smiling, and when he had finished he said, "Let it go that way. It is written in kindness. Such a thing can't be bad."

But it was bad in several ways neither of us foresaw. As the book began to be read, tourists began coming to the laboratory, first a few and then in droves. People stopped their cars and stared at Ed with that glassy look that is used on movie stars. Hundreds of people came into the lab to ask questions and peer around. It became a nuisance to him. But in a way he liked it too. For as he said, "Some of the callers were women and some of the women were very nice looking." However, he was glad when the little flurry of publicity or notoriety was over.

It never occurred to me to ask Ed much about his family background or his life as a boy. I suppose it would be easy to find out. When he was alive there were too many other things to talk about, and now—it doesn't matter. Of course I have heard him asked the usual question about his name. Ricketts. He said, "No, I was not named after the disease—one of my relatives is responsible for its naming."

When the book *Studs Lonigan* came out, Ed read it twice very quickly. "This is a true book," he said. "I was born and grew up in this part of Chicago. I played in these streets. I know them all. I know the people. This is a true book." And, of course, to Ed a thing that was true was beautiful. He followed the whole series of Farrell's books after that and only after the locale moved to New York did he lose interest. He did not know true things about New York.

One of the most amusing things that ever happened in Pacific Biological Laboratories was our attempt to help with the war effort against Japan and the complete fiasco that resulted.

When we came back from the collecting trip which is recorded in the latter part of this book, we went to work on the thousands of animals we had gathered. Our project had been to lay the basis for a new faunal geography rather than a search for new species. We needed a great amount of supplementary information regarding the distribution of species on both sides of the Pacific Ocean and among the Pacific Islands, since many species are widely placed.

By this time Pearl Harbor had been attacked and we were at war with Japan. But even if we had not been, there were difficulties. Soon after the First World War a great number of the islands of the Pacific were mandated to Japan by the League of Nations. And Japan's first act had been to draw a bamboo curtain over these islands and over the whole area. No foreigner had been permitted to land on them in twenty years for any purpose whatever.

These islands had not been well known in a zoologic sense before the mandate and nothing had been heard from them since—so we thought.

We sent out the usual letters to universities, requesting information that might be available concerning these curtained islands. The replies delighted us. There was a great deal of information available.

What had happened is this. Japan had certainly cut off the islands from the world, but, perhaps with the future war in view, Japan had wanted to make a survey of her new possessions in the matter of food supply from the ocean. The Japanese eat many more sea products than we do. Who better to send to make this survey than certain eminent Japanese zoologists who were internationally known?

What followed is truly comic opera. The zoologists did make the survey—very secretly. Then afterwards, since they were good scientists and specialists, what was more natural than that they should study their specialties together with the ecological theater? And then, being thoroughly good men, they completed their zoologic survey.

Now a careful zoologic survey notes not only the animals but their neighbor animals—friends and enemies and the conditions under which they live. Such conditions would include weather, wave shock, tidal range, currents, salinity, reefs, headlands, winds, nature of coast and nature of bottom, and any interesting phenomena which might interfere with or promote the occurrence, normal growth, and happiness of the animals in question. Such matters might be mentioned as the discharge into tide pools of by-products of new chemical plants which would change the ecological balance.

Having finished their sea-food reports to the Japanese government, the zoologists with even more loving care wrote their papers on the specialties. And then, what was more natural than that they should send these papers to their colleagues around the world? Japan was not at war. They knew their brother zoologists would be interested and many of the Japanese had studied at Harvard, at Hopkins, at California Institute of Technology—in fact at all of the American universities. They had friends all over the world who would appreciate and applaud their work in pure science.

When these surveys began to arrive Ed and I suddenly lost our interest in the animals. Here under our hands were detailed studies of the physical make-up of one of the least-known areas of the world and one which was in the hands of our enemy. With excitement we realized that if we were ever going to go island-hopping toward Japan, which seemed reasonable, here was all the information needed if we were to make beach landings—depth, tide, currents, reefs, nature of coast, etc. We did not know whether we were alone in our discovery. We wondered whether our naval or military intelligence knew of the existence of these reports. Often a very obvious thing may lie unnoticed. It seemed to us that if our intelligence services did not know, they should, and we were quite willing to take the chance of duplication.

We drafted a letter to the Navy Department in Washington, explaining the material, its possible use, and how we had come upon it.

Six weeks later we received a form letter thanking us for our patriotism. I seem to remember that the letter was mimeographed. Ed was philosophical about it, but I, who did not have his military experience and cynicism, got mad. I wrote to the Secretary of the Navy, at that time the Honorable Frank Knox, again telling the story of the island material. And then after the letter was sealed, in a moment of angry impudence, I wrote "Personal" on the envelope.

Nothing happened for two months. I was away when it did happen. Ed told me about it later. One afternoon a tight-lipped man in civilian clothes came into the laboratory and identified himself as a lieutenant commander of Naval Intelligence.

"We have had a communication from you," he said sternly.

"Oh, yes," Ed said. "We're glad you are here."

The officer interrupted him. "Do you speak or read Japanese?" he asked suspiciously.

"No, I don't," Ed said.

"Does your partner speak or read Japanese?"

"No—why do you ask?"

"Then what is this information you claim to have about the Pacific islands?"

Only then did Ed understand him. "But they're in English—the papers are all in English!" he cried.

"How in English?"

"The men, the Japanese zoologists, wrote them in English. They had studied here. English is becoming the scientific language of the world."

This thought, Ed said, really made quite a struggle to get in, but it failed.

"Why don't they write in Japanese?" the commander demanded.

"I don't know." Ed was getting tired. "The fact remains that they write English—sometimes quaint English but English."

That word tore it, just as my "Personal" on the envelope probably tore it in Washington.

The lieutenant commander looked grim. "Quaint!" he said. "You will hear from us."

But we never did. And I have always wondered whether they had the information or got it. I wonder whether some of the soldiers whose landing craft grounded a quarter of a mile from the beach and who had to wade ashore under fire had the feeling that bottom and tidal range either were not known or were ignored. I don't know.

Ed shook his head after he told me about the visit of the officer. "I never learn," he said. "I really fell into that one. And I should know better. And I used to be a company clerk." Then he told me about the Navy tests at Bremerton.

The tests were designed to develop some bottom material or paint which would repel barnacles. The outlay of money was considerable—big concrete tanks were built and samples of paints, metal salts, poisons, tars, were immersed to see what material barnacles would be most likely to stay away from.

"Now," Ed said, "a friend of mine who teaches at the University of Washington is one of the world's specialists in barnacles. My friend happens to be a woman. She heard of the tests and offered her services to the Navy. A very patriotic woman as well as a damn good scientist.

"There were two strikes against her," Ed said. "One, she was a woman, and two, she was a professor. The Navy was gallant but adamant. She was thanked and informed that the Navy was not interested in theory. This was hard-boiled realism, and practical men—not theoreticians—would see it through."

Ed grinned at me. "You know," he said, "at the end of three months there wasn't a single barnacle on any of the test materials, not even on the guide materials, the untouched wood and steel. My friend heard of this and visited the station again. She was shy about imposing theory. But she saw what was wrong very quickly.

"The Navy is hard-boiled but it is clean," Ed said. "Bremerton water, on the other hand, is very dirty—you know, harbor stuff, oil and algae, decayed fish and even some human residue. The Navy didn't like that filth so the water was filtered before it went to the tanks. The filters got the water clean," said Ed, "but it also removed all of the barnacle larvae." He laughed. "I wonder whether she ever told them," he said.

Thus was our impertinent attempt to change the techniques of warfare put in its place. But we won.

I became associated in the business of the laboratory in the simplest of ways. A number of years ago Ed had gradually got into debt until the interest on his loan from the bank was bleeding the laboratory like a cat in the basement. Rather sadly he prepared to liquidate the little business and give up his independence—the right to sleep late and work late, the right to make his own decisions. While the lab was not run efficiently, it could make enough to support him, but it could not also pay the bank interest.

At that time I had some money put away and I took up the bank loans and lowered the interest to a vanishing point. I knew the money would vanish anyway. To secure the loan I received stock in the corporation—the most beautiful stock, and the mortgage on the property. I didn't understand much of the transaction but it allowed the laboratory to operate for another ten years. Thus I became a partner in the improbable business. I must say I brought no efficiency to bear on it. The fact that the institution survived at all is a matter that must be put down to magic. I can find no other reasonable explanation. It had no right to survive. A board of directors' meeting differed from any other party only in that there was more beer. A stern business discussion had a way of slipping into a consideration of a unified field hypothesis.

Our trip to the Gulf of Lower California was a marvel of bumbling efficiency. We went where we intended, got what we wanted, and did the work on it. It had been our intention to continue the work with a survey of the Aleutian chain of islands when the war closed that area to us.

At the time of Ed's death our plans were completed, tickets bought, containers and collecting equipment ready for a long collecting trip to the Queen Charlotte Islands, which reach so deep into the Pacific Ocean. There was one deep bay with a long and narrow opening where we thought we might observe some changes in animal forms due to a specialized life and a long period of isolation. Ed was to have started within a month and I was to have joined him there. Maybe someone else will study that little island sea. The light has gone out of it for me.

Now I am coming near to the close of this account. I have not put down Ed's relations with his wives or with his three children. There isn't time, and besides I did not know much about these things.

As I have said, no one who knew Ed will be satisfied with this account. They will have known innumerable other Eds. I imagine that there were as many Eds as there were friends of Ed. And I wonder whether there can be any parallel thinking on his nature and the reason for his impact on the people who knew him. I wonder whether I can make any kind of generalization that would be satisfactory.

I have tried to isolate and inspect the great talent that was in Ed Ricketts, that made him so loved and needed and makes him so missed now that he is dead. Certainly he was an interesting and charming man, but there was some other quality which far exceeded these. I have thought that it might be his ability to receive, to receive anything from anyone, to receive gracefully and thankfully and to make the gift seem very fine. Because of this everyone felt good in giving to Ed—a present, a thought, anything.

Perhaps the most overrated virtue in our list of shoddy virtues is that of giving. Giving builds up the ego of the giver, makes him superior and higher and larger than the receiver. Nearly always, giving is a selfish pleasure, and in many cases it is a downright destructive and evil thing. One has only to remember some of our wolfish financiers who spend two-thirds of their lives clawing fortunes out of the guts of society and the latter third pushing it back. It is not enough to suppose that their philanthropy is a kind of frightened restitution, or that their natures change when they have enough. Such a nature never has enough and natures do not change that readily. I think that the impulse is the same in both cases. For giving can bring the same sense of superiority as getting does, and philanthropy may be another kind of spiritual avarice.

It is so easy to give, so exquisitely rewarding. Receiving, on the other hand, if it be well done, requires a fine balance of self-knowledge and kindness. It requires humility and tact and great understanding of relationships. In receiving you cannot appear, even to yourself, better or stronger or wiser than the giver, although you must be wiser to do it well.

It requires a self-esteem to receive—not self-love but just a pleasant acquaintance and liking for oneself.

Once Ed said to me, "For a very long time I didn't like myself." It was not said in self-pity but simply as an unfortunate fact. "It was a very difficult time," he said, "and very painful. I did not like myself for a number of reasons, some of them valid and some of them pure fancy. I would hate to have to go back to that. Then gradually," he said, "I discovered with surprise and pleasure that a number of people did like me. And I thought, if they can like me, why cannot I like myself? Just thinking it did not do it, but slowly I learned to like myself and then it was all right."

This was not said in self-love in its bad connotation but in self-knowledge. He meant literally that he had learned to accept and like the person "Ed" as he liked other people. It gave him a great advantage. Most people do not like themselves at all. They distrust themselves, put on masks and pomposities. They quarrel and boast and pretend and are jealous because they do not like themselves. But mostly they do not even know themselves well enough to form a true liking. They cannot see themselves well enough to form a true liking, and since we automatically fear and dislike strangers, we fear and dislike our stranger-selves.

Once Ed was able to like himself he was released from the secret prison of self-contempt. Then he did not have to prove superiority any more by any of the ordinary methods, including giving. He could receive and understand and be truly glad, not competitively glad.

Ed's gift for receiving made him a great teacher. Children brought shells to him and gave him information about the shells. And they had to learn before they could tell him.

In conversation you found yourself telling him things—thoughts, conjectures, hypotheses—and you found a pleased surprise at yourself for having arrived at something you were not aware that you could think or know. It gave you such a good sense of participation with him that you could present him with this wonder.

Then Ed would say, "Yes, that's so. That's the way it might be and besides—" and he would illuminate it but not so that he took it away from you. He simply accepted it.

Although his creativeness lay in receiving, that does not mean that he kept things as property. When you had something from him it was not something that was his that he tore away from himself. When you had a thought from him or a piece of music or twenty dollars or a steak dinner, it was not his—it was yours already, and his was only the head and hand that steadied it in position toward you. For this reason no one was ever cut off from him. Association with him was deep participation with him, never competition.

I wish we could all be so. If we could learn even a little to like ourselves, maybe our cruelties and angers might melt away. Maybe we would not have to hurt one another just to keep our ego-chins above water.

There it is. That's all I can set down about Ed Ricketts. I don't know whether any clear picture has emerged. Thinking back and remembering has not done what I hoped it might. It has not laid the ghost.

The picture that remains is a haunting one. It is the time just before dusk. I can see Ed finishing his work in the laboratory. He covers his instruments and puts his papers away. He rolls down the sleeves of his wool shirt and puts on his old brown coat. I see him go out and get in his beat-up old car and slowly drive away in the evening.

I guess I'll have that with me all my life.

GLOSSARY

OF TERMS AS USED IN THIS WORK

ABORAL. The upper surface of a starfish, brittle-star, or sea-urchin, as opposed to the under or oral surface whereon the mouth is situated.

ALGAE. Simple plants, often unicellular; the higher forms include the seaweeds.

AMBULACRAL GROOVE. A furrow bisecting the underside of the rays of starfish through which the tube feet are protruded.

AMPHIPOD. Literally, "paired-legs." Minute shrimp-like crustaceans, laterally compressed; the beach hoppers, sand fleas, skeleton shrimps, etc.

ANASTOMOSING. Dictionary definition: "Union or intercommunication of any system or network of lines, branches, streams, or the like."

ASSOCIATION. An assemblage of animals having ecologically similar requirements.

ATOKOUS. The sexually immature stage of certain polychaet worms.

AUTONOMY. Reflex, or seemingly voluntary, separation of a part or a limb from the body, followed by regeneration.

BUNODID ANEMONE. One of a family of sea-anemones characterized by a bumpy or warty body wall.

CALCAREOUS. Containing deposits of calcium carbonate; calcification.

CERATA. Dorsal projections which take the place of gills.

COMMENSAL. An organism living in, with, or on another, generally partaking of the same food.

COSINE WAVE. A wave graphically represented by a curving line, the peaks and troughs of which are equal and complementary.

CTENOPHORE. A type of jellyfish characterized by the possession of meridional rows of vibrating plates which propel and orient the animal.

DACTYL. Term applied to the last joint of a crustacean leg.

DEHISCENCE. A bursting discharge, usually of eggs or sperm.

DROWNED CORAL FLAT. A flat containing coral, some heads of which have been suffocated by sand.

ECHIUROID. A worm-like animal related to the sipunculids, in which the body is variably sac-like, usually with thin skin, and having often a spoon-shaped proboscis.

ECOLOGY. The study of the mutual relations between an organism and its physical and sociological environment.

ELYTRA. Shield-like scales of certain worms.

ENDEMIC. Dictionary example: "An *endemic disease* is one which is constantly present to a greater or less degree in any place, as distinguished from an *epidemic disease*, which prevails widely at some time, or periodically. . . ."

EPITOKOUS. Sexually mature stage in polychaet worms, characterized by changes of the posterior end which enable normally crawling worms to be free-swimming.

ETIOLOGY. Dictionary definition: "1. The science, doctrine, or demonstration of causes, especially the investigation of the causes of any disease. 2. The assignment of a cause or reason; as, the *etiology* of a historical custom."

FLORIATE. Flower-like.

GASTROPOD. Literally, "stomach-foot." Belonging to a group of animals comprising the snails, slugs, sea-hares, etc.

GYMNOBLAST. Belonging to a group (of hydroids) in which the polyps lack the skeletal cups of other hydroids into which the soft parts can be withdrawn.

HOLOTHURIAN. Sea-cucumber. One of a group of echinoderms, or spiny-skinned animals, some varieties of which, under the commercial name *bêche-de-mer* or *trepang*, are used by the Chinese for food.

HYDROID. A small, plant-like, usually colonial animal.

INTERTIDAL. See *Littoral*.

INTROVERT. A closed tubular pocket capable of being unrolled and extended inside out.

ISOPOD. Literally, "same legs." Usually small crustaceans in which all the legs are similar, comprising the pill-bugs, sow-bugs, and many marine forms.

ISOTHERM. A line joining or marking equal temperatures.

LITTORAL. Region of the shore bounded by its highest normal submergence at high tide and most extreme emergence at low tide. Intertidal.

MUTATION. In the life history of a species, the sudden appearance of a new trait that breeds true and becomes eventually one of the characters of the species or of the new species thus formed.

MYSIDS. Usually minute crustacea, called "opossum shrimps" because of their possession of marsupial plates within which the young develop.

NUDIBRANCH. Literally, "naked gill." One of a group of shell-less gastropods, often brilliantly colored and of delicately beautiful form.

OPHIURAN. Brittle-star or serpent-star. Members of one of the five classes of echinoderms or spiny-skinned animals.

PAPILLA. Small elevation; in holothurians, modified tube feet not used for locomotion.

PELAGIC. Free-floating at or near the surface of the sea.

PLANKTON. Generally microscopic plant and animal life floating or weakly swimming in the upper layer of a body of water.

POLYCHAETS. Usually elongate worms characterized by the possession of abundant chaetae or bristles.

POLYCLADS. Flatworms in which the intestinal tract has extensive ramifications.

POLYP. An invertebrate having a hollow cylindrical body, closed and attached at one end and opening at the other by a central mouth surrounded by tentacles. May be an individual (as an anemone) or a member of a colony (as a coral polyp).

PORCELLANIDS. Crabs of the family Porcellanidae, often called porcelain crabs because of the carapace texture of typical examples.

QUATERNARY, OR RECENT. The latest of the epochs into which geologists divide the history of the earth. Late Quaternary includes the present time.

RESPIRATORY TREE. The respiratory organ of holothurians; so named because it resembles a tree inside out. Fresh water is taken in at what corresponds to the trunk and penetrates to the delicate branches, which provide great absorption area in proportion to the volume.

SCALAR. Mathematical term. An abstract quantity having magnitude but not direction, such as volume, mass, weight, time, electrical charge, and always indicated by a real number.

SERPULID. A polychaet worm which builds a calcareous tube, usually coiled.

SESSILE. Attached, therefore not moving.

SIPHONOPHORE. A type of jellyfish. The Portuguese man-o'-war and other spectacular forms belong to this group.

SIPUNCULIDS. Worm-like animals characterized (among other things) by the possession of an introvert, and of rough, cuticle-like skin. Capable of great expansion; contracted, some of them merit the name peanut worm.

SYNDROME. A group of signs and symptoms occurring together and characterizing a disease.

SYNONYMY. The various names used to designate a given species or group.

TAXONOMY. A sub-science of biology concerned with the classification of animals according to natural relationships and with the rules governing the system of nomenclature.

TECTIBRANCHS. A group of sometimes shell-less gastropods to which belong the sea-hares and bubble-shells.

TELEOLOGY. The assumption of predetermined design, purpose, or ends in Nature by which an explanation of phenomena is postulated.

TENSOR. A mathematical term for the stretching factor which is necessary to change one vector, or force, into another vector having a different amount of force and direction. (Thus, if one imagines a given force *A* traveling south at 40 miles an hour, and another force *B* traveling southeast at 60 miles an hour, mathematically to translate force *A* into force *B*, the factor which changes one into the other must have not only force and direction, but stretching power, to pull *A* equal to *B*, and that factor is called the *tensor*.) Tensor is the quantity necessary in Einsteinian physics to translate vectors from one set of co-ordinates (frame of reference) to another.

TEREBELLID WORM. A polychaet worm which builds a sandy or pebbly tube, cemented usually to the underside of rocks by its own mucus.

THIGMOTROPISM. An innate tendency to seek enclosing contact with a solid or rigid surface, as in a burrow.

TROPISM. Innate involuntary movement of an organism or any of its parts toward (positive) or away from (negative) a stimulus.

TURBELLARIAN WORMS. The large group of flatworms to which the polyclads belong.

UBIQUITOUS. Occurring everywhere (though not necessarily abundantly) in the total area under consideration.

VECTOR. A mathematical term for an abstract quantity such as velocity, acceleration, or force, having *both* magnitude and direction. It may also have position in space, but this is not necessary. A vector is symbolized or represented by an arrow.

XEROPHYTIC. Plants structurally adapted to withstand drought.

ZOOID. Individual member of a colony or compound organism, having more or less independent life of its own.

INDEX